BASIC BENCHWORK FOR HOME MACHINISTS

BASIC BENCHWORK FOR HOME MACHINISTS

Les Oldridge

© Les Oldridge 1988
Published in the UK by Special Interest Model Books 2013
This edition published in 2020 by Fox Chapel Publishing Company, Inc., 903 Square Street, Mount Joy, PA 17552.

Fox Chapel Publishing Edition
Technical Editor: George Bulliss
Editor: Anthony Regolino
Layout: Christopher Morrison

ISBN 978-1-4971-0057-2

Library of Congress Control Number:2019950252

To learn more about the other great books from Fox Chapel Publishing, or to find a retailer near you, call toll-free 1-800-457-9112 or visit us at *www.FoxChapelPublishing.com*.

We are always looking for talented authors. To submit an idea, please send a brief inquiry to acquisitions@foxchapelpublishing.com.

Printed in Singapore
First printing

Contents

Chapter 1

Introduction

Modern engineering workshops are equipped with machine tools capable of producing components to such accurate limits that hand fitting at the bench is no longer necessary. Mass production methods render the skills possessed by old-time fitters in danger of being forgotten forever. This is a pity, as in many situations, the ability to complete a job using only hand tools is a great asset. In any case, it is often quicker to bring a component to the correct dimensions using a hand method than it is to spend time setting up the job in a milling machine or shaper, even if one is available.

Machine tools owned by the model engineer are often limited to a lathe and, perhaps, a bench drilling machine, so he has to become skilled in the use of hand tools. The purpose of this book is to describe the basic skills he must acquire. It takes a great deal of practice to reach the standard required, and although disappointment may be experienced at first, with the slow progress made, the satisfaction when the job is concluded is well worth the work involved.

It must be emphasized that no textbook, however comprehensive, can take the place of actual experience at the bench. The best advice is to "have a go," perhaps on a bit of scrap material, to gain the necessary skill and confidence before working on a valuable casting.

Throughout this book, special emphasis will be made on safe working. In industry, this is looked after by the Health and Safety at Work Executive, but in the amateur's workshop, there is no legislation to ensure the worker has a safe working environment. It is up to the individual to look after himself and to help to this end the various hazards likely to be encountered will be outlined from time to time.

A simple first aid kit, and the knowledge of how to use it, is desirable. A fire extinguisher is also a good investment and the Fire Prevention Officer from the local fire department will be pleased to give free advice as to the best type to purchase to suit your particular needs.

If good work is to be produced, a sturdy, rigid bench, fitted with a good quality vise, is essential. Most model engineers are not wealthy and setting up a workshop is a costly business. I hope to suggest, where possible, ways of cutting costs without sacrificing quality. For example, good secondhand timber is often available at low prices; when buildings are being demolished, a tactful word with the site

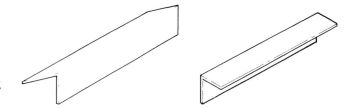

Fig. 1.1 *Vise clamps bent up in aluminum, copper, or lead, dimensions to suit vise.*

foreman may provide just what is needed to build a bench at low cost.

Legs made from at least 3 in. square timber and the top from 2 in. planks should be aimed at. The space between the legs can be used to house a useful cupboard.

The size of the vise will depend on the type of work to be undertaken, but it is better to have one a little larger than is thought to be necessary, to allow for "future expansion," that is, for the bigger jobs which may come along later.

The height of the bench should be such that the top of the vise jaws are in line with the point of the user's elbow. This makes filing accurately much easier, but more about that later.

The vise jaws are serrated to prevent the work slipping when roughing down. Clamps, sometimes called clams, to fit over the jaws are needed to prevent these serrations from damaging finished surfaces. They may be made of lead, copper, aluminum, or fiber or any other soft material (see Fig. 1.1).

Various special clamps can be made to hold round, or odd, shaped work securely in the vise. Fig. 1.2. shows an easily made and useful device for holding round bar or pipe. Holes of a size to suit the bars in common use are drilled, as shown, in a piece of mild steel 25mm × 12mm (½ in. × 1 in.) and of a length to accommodate the number of holes required. A saw cut is made through the center of all the holes, except the end one. A similar gadget for holding threaded material can be made in a like fashion, except that the holes are threaded before the gadget is split with the type of thread generally used. These two projects form a useful exercise after reading the chapters on hacksawing, drilling, and cutting screw threads!

HOLE SIZES TO SUIT MARKER'S REQUIREMENTS

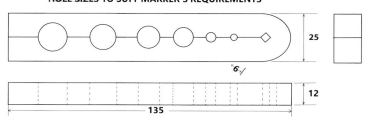

Fig. 1.2 *Vise clamp for holding round bars.*

DIMENSIONS IN MILLIMETERS. MATERIAL IN BRIGHT MILD STEEL

Chapter 2

Materials

Before looking at the various tasks which are performed at the bench, the materials on which we shall be working and their properties must be discussed. It is important that the most suitable material for the job in hand is chosen. Often this will be specified in the drawing from which we are working, but sometimes we have to decide what to use.

The following properties then have to be considered:

STRENGTH. The strength of a material is its ability to withstand stress without breaking. The load, or stress, may tend to stretch, compress, twist, or cut the material. These are termed tensile, compressive, torsional, or shear forces. See Fig. 2.1. The strength of a material varies with the type of stress to which it is subject. For example, cast iron has good compressive strength but relatively poor tensile strength; it is about four times stronger when it is squeezed than when it is stretched.

ELASTICITY is the ability of a stressed material to return to its original shape when the load is removed. Spring steel has a high elasticity factor. Plasticine has practically no elasticity. Most materials are elastic below a certain limit, known as their elastic limit. If the stress applied exceeds this limit, the material is permanently deformed.

PLASTICITY is the reverse of elasticity and is the property of a material to retain any deformation produced by loads after the load has been removed. Steel is plastic at red heat and can be forged to shape.

DUCTILITY is the ability in a material to be drawn out by tensile forces beyond its elastic limit without breaking. This property is important in the production of wire, the wire being produced by drawing metal through dies that get progressively smaller.

MALLEABILITY is a similar property to ductility, except that the material is deformed beyond the elastic limit by compressive forces, such as rolling or hammering, instead of by a tensile force. Lead is a malleable material but lacks ductility because of low tensile strength.

BRITTLENESS. A material is brittle where fractures occur with little or no deformation. Glass is a classic example of a material with this property.

TOUGHNESS is the ability to withstand shock loads.

HARDNESS is the ability of a material to resist penetration, scratching, abrasion, indentation, and wear. In the laboratory,

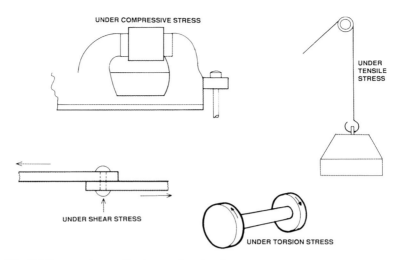

Fig. 2.1 *Compressive, tensile, shear, and torsional stresses.*

it is measured by applying a load to a small area of material by a hard steel ball or pointed diamond, and measuring the depression made into the material under a given load. Chisels, lathe tools, and center punches, for example, must have this quality to do the job for which they are intended. Unfortunately, the harder carbon steel tools are made the more brittle they become, so some hardness must be sacrificed for toughness in the tempering process. This will be discussed more fully in the chapter on hardening and tempering.

SOFTNESS, obviously, is the opposite property to hardness. Soft materials may be easily shaped by filing, drilling, or machining in a lathe, milling machine, or shaper. In many cases the component is hardened by one means or another, to be discussed later, after the shaping process is completed.

MATERIALS

Materials can be divided into a number of groups, such as:

1. Metals, which can be subdivided into ferrous and non-ferrous metals. This is the group with which we are most concerned but the others will be met from time to time.
2. Plastics, are now widely used in industry and which the model engineer will occasionally use them.
3. Timber.
4. Ceramics—the name originally given to materials made from clay but now used to cover a wide range of materials.

FERROUS METALS

These are the metals containing iron. Metals are rarely used in their pure state but are combined with other metals to form an **ALLOY**. In the case of iron, carbon is the most important addition. Although it is only present in small amounts, it causes big changes in the property of the metal.

CAST IRON this form the iron has been melted and poured into a mold, usually made of sand, in which it is allowed to solidify. This is a simple, convenient, and relatively cheap process to manufacture components of a complicated shape. Cast iron is an alloy of iron and carbon with small amounts of manganese, silicon, sulfur, and phosphorus. It contains about 3% of carbon.

There are two types, gray and white. Both get their names from the appearance of the metal when fractured. In white cast iron, all the carbon present is cementite; in gray cast iron, most of the carbon is present as flakes of graphite, and there is usually a remainder which is in the form of pearlite. Because cementite is intensely hard, white cast iron is hard and durable, though very brittle. Graphite is soft and is a good lubricant, so gray cast iron is readily machinable, less brittle, and suitable for sliding surfaces. Being hard and brittle white cast iron is rarely used alone but it is the material used for the production of malleable iron.

GRAY CAST IRON, then, is the type in common use; it is cheap and easy to cast and machine. As a typical example, a motor car cylinder block contains 93.32% iron, 3.3% carbon, 1.9% silicon, 0.8% manganese, 0.14% sulfur, and 0.18% each of phosphorus, molybdenum, and chromium. The carbon content of approximately 3.3% consists of about 0.7% of combined carbon and about 2.6% of free carbon.

Because of the free carbon content, cast iron is easy to machine and file; the carbon flakes act as a lubricant, enabling the cast iron to be machined dry. Drilling or tapping of cast iron components is fairly easy, no lubricant being required. There is, however, a hard skin in which some of the molding sand may still be present. This is particularly hard on lathe tools, and when it has to be filed, an old file should be used; a new one would probably be ruined.

Cast iron is used for model engine flywheels, internal combustion engine cylinders, model locomotive wheels, and a host of other parts. Because of its self-lubricating properties, it is an ideal material for plummer block bearings. The spindle of the *Model Engineer* sensitive drilling machine runs directly in cast iron bearings and shows little signs of wear after years of use.

Cast iron has low tensile strength and poor shock resistance.

THE STEELS

There are standard specifications for steels contained in BS 970, which dates back to 1942, but since that date, there have been several revisions. In 1970, the specifications underwent a radical change and in 1983 the Standards were again restructured. Originally an EN code was used but this is now replaced by a six-digit system. It will be some time before the EN numbers disappear altogether and, in fact, some manufacturers show both the EN numbers and the current specifications where the two are closely aligned with only a point or two variation in analysis. For example, the free-cutting steel 212M36 corresponds to the old EN8M.

The North American naming convention uses a four digit number that helps to identify the steel type and its carbon content. Common steel types used by hobbyists include 1018, a common, low carbon steel, and 12L14, a low carbon steel with lead added to improve its machinability.

PLAIN CARBON STEELS. The main difference between cast iron and steel is the carbon content. Plain carbon steel has never more than 1.5% carbon, whereas cast iron, as has been stated above, has about 3%.

MILD STEEL, containing about 0.15% to 0.3% carbon combined with the iron, is ductile and malleable. It is easy to weld, machine, forge, or press into a new shape. It may be worked hot or cold. Because of its low carbon content, it cannot be hardened by heating and quenching, but can be case-hardened, a process which will be described later. It is supplied in bar form with hexagon, round, square, or flat sections in a "black" or "bright" form, and in sheets of varying thicknesses.

MEDIUM CARBON STEEL, with a carbon content of 0.35% to 0.5%, is much stronger than mild steel. Its hardness and strength can be increased by quenching the metal from a red heat. It can be tempered, rendering it suitable for many general engineering purposes where the stresses imposed are greater than could be withstood by mild steel.

HIGH CARBON STEEL, with a carbon content of 0.55% to 1.5%, is used for most tools after being hardened and tempered. Chisels, files, drills, and reamers are made from this steel.

ALLOY STEELS. In order to improve the properties of steel and to suit the metal to special applications, other substances beside carbon are added to the steel. **NICKEL** improves the ductility and toughness of the metal. **CHROMIUM** and **MOLYBDENUM** increase its hardness, while **VANADIUM** improves the elasticity, strength, and fatigue resistance of the steel. All steel contains **MANGANESE** but sometimes more is added to improve the steel's mechanical properties.

STAINLESS STEEL is principally an alloy of iron, nickel, and chromium. It has a high resistance to corrosion, but in some forms, it is difficult to machine. However, by introducing a free machining agent into the alloy, this drawback can be overcome. **DRILL ROD**, a common tool steel used by model engineers, is a carbon steel with 1.1% to 1.2% carbon, 0.35% manganese, 0.45% chromium, and 0.1% to 0.25% of silicon.

TINPLATE. Sheets of mild steel are coated with tin to provide the metal used for the familiar food containers and for many other purposes. It is a useful material for the model engineer, being easily worked and soldered and can be obtained without cost from discarded cookie tins, etc.

NON FERROUS METALS

ALUMINUM is the lightest of the commonly used metals. It is too soft to use in its pure state, but when alloyed with copper, magnesium, and manganese, it is widely used for many components. It is a good conductor of electricity but is impossible to solder by the usual methods.

COPPER is soft, ductile, and of low tensile strength. It is an excellent conductor of electricity and is easy to solder or braze. It is the base of the brass and bronze alloys. Copper hardens with age and also work-hardens, that is, it becomes hard when it is bent or stretched. It can easily be returned to its soft, ductile state by annealing. This is done by heating to a red color and then allowing it to cool.

LEAD is soft, ductile and of very low tensile strength. It is often added to other metals to make them free cutting. It is typically used for lead acid battery plates and in soft solder.

TIN is corrosion resistant and is used to coat mild steel plate to make "tin plate." It is used in soft solder and is an alloying agent in bronze, and is the basis of "white metal" bearings.

BRASS AND BRONZE. When copper is alloyed with zinc, brass is formed. Bronze is an alloy of copper and tin, and usually about 10% tin is used. Sometimes about 0.5% of phosphorus is added, the alloy then being called phosphorous bronze.

There are various classes of bronze made especially for particular applications. It is, for example, an excellent bearing material.

IDENTIFICATION OF FERROUS METALS

Several metals have a similar appearance and new bar materials are often color coded by painting the end with a distinctive color paint. Often, off-cuts are used up and it is essential that these are identified. Most model engineers have a scrap box where all sorts of odds and ends are stored. Trouble will be experienced if, for example, a piece of high carbon steel is selected when a free cutting mild steel is what is required. There are several ways in which metals can be identified and some tests appear below.

APPEARANCE. Cast iron has a dark rough finish; the mold joint line is probably visible. A section of iron away from the skin has a gray appearance and a fracture appears crystallized.

Mild steel comes in two forms, black and bright. The former, "hot-rolled steel," has a smooth scale with a blue/black sheen. Cold-rolled steel (CRS) has a bright, silver-gray surface. Medium carbon steel has a smooth scale and a black sheen, while high carbon steel has a rougher scale.

GRINDING. A popular test is to grind the metal and note the color, quantity, and type of sparks given off. This is a difficult procedure to describe: a video film is really necessary, and it is equally difficult for the beginner to recognize the different types of sparks. It is suggested that an experiment

12

is carried out using steels of known types and comparing the differences.

Cast iron gives off a short stream of red sparks, which at some distance from the grinding wheel, burst into a yellow spark formation. Plain carbon steel produces a lighter and brighter spark in a greater profusion than cast iron. As the carbon content increases, the sparks become lighter, are in greater quantities, and occur nearer the wheel. The high carbon steels produce secondary bursts, bunching out from the primary sparks.

If materials are drilled, it is very noticeable that the cuttings from cast iron are granular in form, while those from steel come off in long spirals. The cuttings (swarf) from medium carbon steel may turn brown or blue, but still be in spiral form. Swarf is very sharp and can cause nasty cuts if handled, so special care is needed when clearing the cuttings away from a drill.

PLASTICS

These materials become plastic above certain temperatures and, while plastic, they can be squeezed into dies or molds to give them the required shape that they retain on cooling. There are two main types, **THERMOSETTING** and **THERMOPLASTIC**. The former group do not become plastic on re-heating. They are hard, rigid, and rather brittle. They are used particularly for electrical equipment as they are good insulators. Bakelite comes within this category.

Thermoplastics may be softened by heat so they cannot be used at temperatures much above 212°F (100°C). Some of them, celluloid and plexiglass (Perspex) for example, are transparent and most can be colored by adding a suitable pigment.

POLYVINYLCHLORIDE (PVC) comes in this class, and is the flexible and rubberlike substance commonly used for insulating electric cables.

POLYTETRAFLUORETHYLENE (PTFE) is similar to PVC but has a very low co-efficient of friction, which makes it particularly suitable for making bushes which need not be lubricated. **NYLON**, one of the earliest plastics, is used for a variety of purposes, including small gearwheels.

REINFORCED PLASTIC. Laminated plastic such as **TUFNOL** consists of a fibrous material like paper or woven cloth impregnated with phenolic resin. The sheets of fabric are then laid up in a hydraulic press and squeezed and heated so that they become solid sheets, rods, or tubes.

GLASS FIBERS can be bonded together by polyester or epoxy resins to form large and complex moldings. Crash helmets and boat hulls are examples of things made in this way. The customary term is "glass-reinforced plastic" or GRP.

Chapter 3

Reading Engineering Drawings

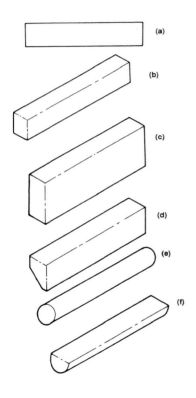

A fitter working at a bench will normally be required to produce components to the dimensions and outline as shown on an engineering drawing so, quite obviously, he must be able to read drawings of this kind. Although it is not intended to study this complex subject at any great depth, certain fundamentals will be explained, sufficiently enough it is hoped, to allow the hobbyist to interpret drawings that he has to work from.

While the main details of a simple part can be shown in a pictorial view, for sufficient information to be available to construct the average component, several related views showing the front, sides, top, and/or bottom are required. For example, the single drawing shown at the top of Fig. 3.1 could be an illustration of any of the shapes shown below.

For all the details required to make all but the simplest of components, drawings using **ORTHOGRAPHIC PROJECTION** are needed. There are two systems: **FIRST ANGLE**, widely practiced in the UK and in North America, and **THIRD ANGLE**, mainly used in North America but occasionally in Britain.

Orthographic projection results when the outline of an object is projected, at right angles, on to a flat surface known as

Fig. 3.1 *View (a) could be any of the other shapes viewed from above.*

14

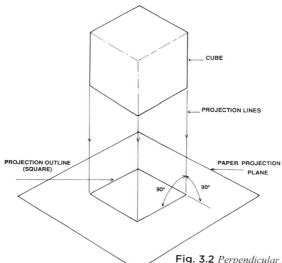

CUBE

PROJECTION LINES

PROJECTION OUTLINE
(SQUARE)

PAPER PROJECTION
PLANE

90°   90°

Fig. 3.2 *Perpendicular orthographic projection.*

a plane. See Fig. 3.2 where lines have been projected from a cube, downwards onto a flat surface.

The two systems of first and third angle take their names from the first and third quadrants of a circle (see Fig. 3.3). Here four open ended "boxes" are shown around the quadrants of a circle. Objects to be drawn are imagined to be placed in one of these boxes and their outlines projected on to the "walls" of the box. First and third angle projections have similar merits; only local preference makes one more popular than the other.

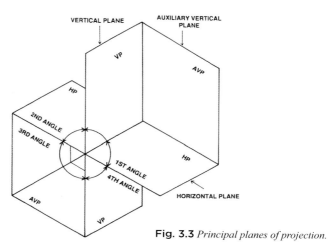

VERTICAL PLANE   AUXILIARY VERTICAL
PLANE

VP

AVP

HP

2ND ANGLE

3RD ANGLE

HP

1ST ANGLE

4TH ANGLE

HORIZONTAL PLANE

AVP

VP

Fig. 3.3 *Principal planes of projection.*

15

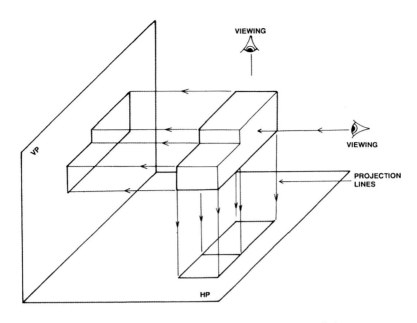

Fig. 3.4 *First angle projection onto vertical and horizontal planes.*

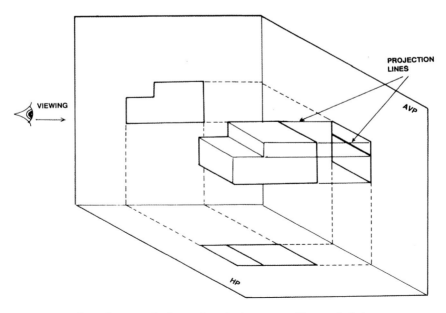

Fig. 3.5 *Viewing for first angle projection onto auxiliary vertical plane.*

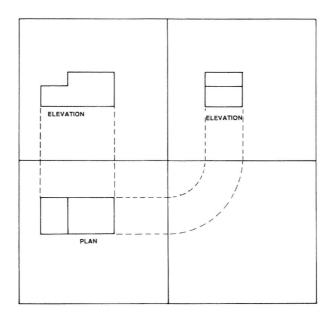

Fig. 3.6 *When "box" is opened out, drawing in first angle orthographic projection will look like this.*

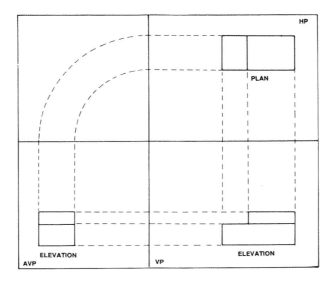

Fig. 3.7 *The same component as the previous, now drawn in third angle projection.*

17

FIRST ANGLE ORTHOGRAPHIC PROJECTION. In Fig. 3.4, perpendicular lines have been projected from an object onto the horizontal and vertical planes in a first angle situation. If the horizontal plane is swung down to the vertical plane, two orthographic views are obtained, an elevation and a plan. A third outline can be viewed and projected on to an auxiliary vertical plane (AVP) as shown in Fig. 3.5. If this plane is swung through 90°, so that it is in line with the other vertical view, the three together give a plan and two elevations of the object, as shown in Fig. 3.6. Generally, these three views will give sufficient detail for a component to be made, but there is no reason why other views (up to six) cannot be projected in the same way to disclose further details.

THIRD ANGLE ORTHOGRAPHIC PROJECTION. Going back to Fig. 3.3, if lines are projected from an object placed in the third angle position onto the horizontal and vertical planes, outlines will appear exactly as viewed. If the sides of the box are then straightened out, as was done in the first angle exercise, then two elevations and a plan will result, as shown in Fig. 3.7. It will be seen that the plan view now appears in the top right section and the two elevations in the lower half of the paper.

Both systems of projection have their own international symbols by which their use can be recognized. The symbols are similar to two views of the frustum of a cone (see Fig. 3.8a and b).

CONVENTIONAL SYMBOLS. In more leisurely times, when labor was cheap, engineering drawings were completed in great detail, every thread on every bolt was shown, every rivet was drawn, but now drawings are very much simplified to save the draftsman's time. Conventional symbols are used for common features such as bolts, studs, and equally spaced holes. Those which are likely to concern benchworkers appear in Fig. 3.9.

HIDDEN DETAILS are shown by short dashes. Fig. 3.10a shows holes in a bar indicated in this way. This sketch also shows how, when a component is too long to accommodate on the paper but is of uniform section, an artificial break is indicated.

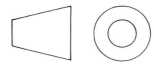

Fig. 3.8a *Symbol for first angle projection drawing.*

Fig. 3.8b *Third angle projection drawing symbol.*

TITLE	SUBJECT	CONVENTION

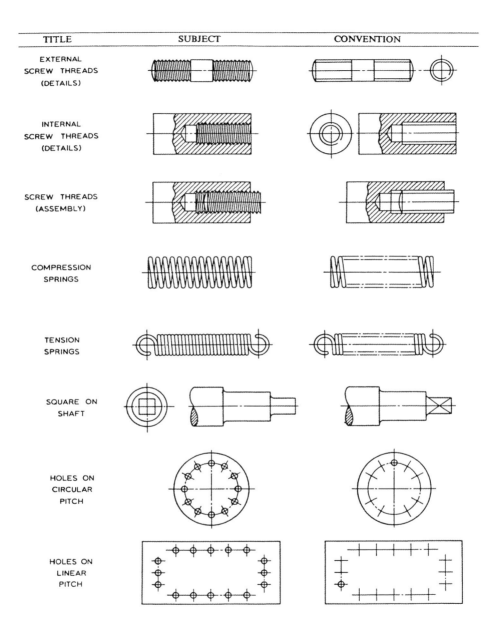

EXTERNAL SCREW THREADS (DETAILS)

INTERNAL SCREW THREADS (DETAILS)

SCREW THREADS (ASSEMBLY)

COMPRESSION SPRINGS

TENSION SPRINGS

SQUARE ON SHAFT

HOLES ON CIRCULAR PITCH

HOLES ON LINEAR PITCH

Fig. 3.9 *Conventional symbols.*

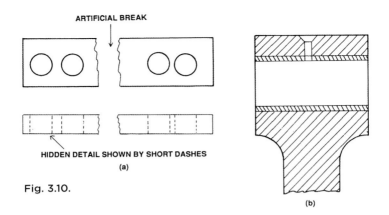

ARTIFICIAL BREAK

HIDDEN DETAIL SHOWN BY SHORT DASHES

(a)

Fig. 3.10.

(b)

SECTIONED VIEWS are views in which it is imagined that the components have been cut through to show their internal construction. Such views are indicated by diagonal lines called hatching lines. Fig. 3.10b shows a connecting rod small end showing the bush, probably made of phosphor-bronze, and the hole for lubricating purposes. The hatching lines are generally at 45° to the centerline of the component. On sectioned features that are adjacent to one another, the hatching lines are drawn at opposite angles.

Sometimes it is necessary for a drawing to contain a special sectional view of a component to make its shape abundantly

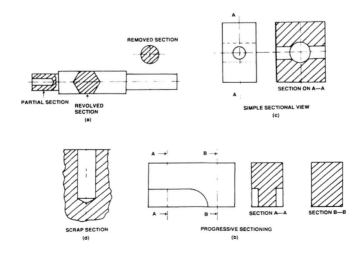

REMOVED SECTION

PARTIAL SECTION   REVOLVED SECTION

(a)

SECTION ON A—A

SIMPLE SECTIONAL VIEW

(c)

SCRAP SECTION

(d)

SECTION A—A   SECTION B—B

PROGRESSIVE SECTIONING

(b)

Fig. 3.11 *Various sectional views.*

20

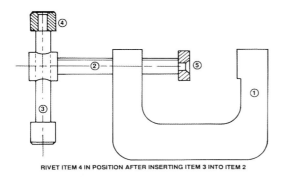

Items to be riveted in position after screwing item 2 in position; allowing item 3 to turn freely on item 2

ITEM NUMBER	DESCRIPTION
1	FRAME
2	SCREW
3	HANDLE
4	FERRULE
5	PAD

RIVET ITEM 4 IN POSITION AFTER INSERTING ITEM 3 INTO ITEM 2

Fig. 3.12 *Assembly drawing of a C-clamp.*

clear; several examples are given in Fig. 3.11. Illustrations (b) and (c) show sectional views where the direction of viewing is shown by arrows. There are several examples of centerlines in these drawings and it will be seen that they consist of a chain line, which is a long thin line followed by a short one.

Usually a component is shown in an **ASSEMBLY DRAWING**, where it is depicted with all its pieces assembled, and a **DETAILED DRAWING** where each single part is shown. An assembly drawing of a C-clamp is shown in Fig. 3.12 and a detailed drawing of the same component in Fig. 3.13.

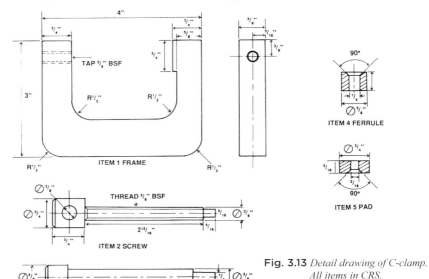

Fig. 3.13 *Detail drawing of C-clamp. All items in CRS.*

21

Chapter 4

Hacksaws

You will often have to cut off a piece of metal to a prescribed length using a hacksaw. The hacksaw, then, is an essential tool in an engineer's workshop, but it is often neglected, misused, or used with a blade that is not correct for the particular job in hand.

Hacksaw blades are classified by their degree of flexibility and how many teeth they have to the inch. High speed steel "all hard" blades are designed to give the optimum cutting performance on all types of materials but they are easily broken if misused. A famous maker describes them as the professional blade suitable for the man who takes a pride in his work and knows exactly how to use them. They can only be used where the workpiece is firmly held.

Probably more suitable for the model engineer is the bi-metal high speed blade, which consists of a high speed cutting edge electron beam welded to a tough alloy back. The makers claim that this blade is virtually unbreakable in normal use and is, therefore, safe in less experienced hands.

There are also flexible blades particularly suitable for work by trainees and occasional use by unskilled workers.

The number of teeth per inch on blades varies from fourteen to thirty-two. The expense involved will preclude the model engineer from stocking the whole range, and some with eighteen and some with twenty-four teeth per inch will probably meet most of his needs.

The number of teeth on the blade to be used depends on the material to be cut and its thickness. As a general rule, three consecutive teeth must be in contact with material being cut. If thin tubing or sheet has to be cut, then a blade with thirty-two teeth per inch is desirable.

The blade must be fitted to the frame with the teeth pointing forward, that is, away from the handle, see Fig. 4.1. The

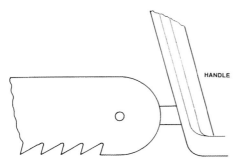

Fig. 4.1 *Teeth must point forward, away from handle.*

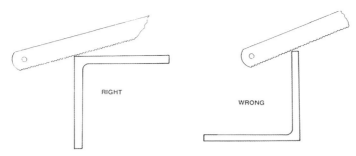

Fig. 4.2 *Cut should be started with as many teeth as possible in contact.*

tensioning wing nut must be turned to take up the "slack," and then tightened three full turns. Correct tension of the blade is important. If the blade is loose, it will buckle and not cut straight; if it is too tight, an unnecessary load is placed on the blade ends and damage to the blade or the frame may result.

The metal to be cut must be securely fastened in the vise. When starting, the cut the thumb of the left hand can act as a guide to position the blade, using very short strokes with light pressure being applied. Do not start the cut on the short edge of the workpiece, but allow as many teeth as possible to be in contact with the work. Fig. 4.2 shows the right and wrong way to start a cut on a piece of angle iron. The wrong method might result in stripped teeth. When the blade has entered the work, the left hand is transferred to the front of the frame and long, steady strokes applied.

Fig. 4.3 and Fig. 4.4 show the incorrect and correct methods of holding work in the vise so that as many teeth as possible are in contact with the work.

Sixty strokes a minute is the correct speed for low speed alloy blades and seventy for high speed ones. Nothing is gained by increasing the speed: strokes using the full length of the blade give the best results and are less tiring. The beginner is very inclined to saw much too fast.

A new blade should not be used in a cut previously made by a worn one, as the old cut will be narrower than the one that the new blade will cut. If a new blade is used in the existing cut, it will jam and could easily break. If a blade breaks partly through a cut, the work should be turned over and the cut started from the other side.

Care is necessary when the saw reaches the end of its cut. Pressure and the rate of stroke should be reduced. Neglecting this precaution can result in injury to the hands and a broken blade.

Keeping a cut straight will come with practice; not too much pressure, a correctly tensioned blade and long, steady strokes all make for accurate work. Cutting carefully to a scribed line will save a lot of tedious filing necessary to bring a job to the prescribed dimension, and to correct the error made in the sawing.

Fig. 4.3 *INCORRECT.*
Work held in vise so that
narrow side is in contact with
the hacksaw teeth.

Hacksaw blades can be obtained in several different lengths and a 12 in. blade is recommended for general use. Hacksaw frames are generally adjustable to allow different length blades to be used.

Junior hacksaws, a small edition of the one already described, are often used by the model engineer for small, delicate jobs. Light and easy to handle, they form a useful addition to the tool kit.

As is common with all types of saws, the teeth of hacksaws are set to the sides. This causes the blade to cut a slot wider than itself, and prevents the body of the blade rubbing or jamming in the saw cut. Often, alternate teeth are set to right and left, every third or fifth tooth being left straight to break up the chips and help the teeth to clear themselves.

Fig. 4.4 *CORRECT.*
Wide side of work
presented to the
hacksaw teeth.

Chapter 5

Files and Filing

Filing is a skill which must be developed by the model engineer, but, unfortunately, a lot of practice is required before the average beginner can become really proficient and be able to file a flat surface. Nevertheless, once the knack has been acquired, the operation becomes automatic and one files flat without thinking about it.

Files are classified according to their length, section, and cut. There are many different types but the model engineer only requires a few. Exactly what he needs depends on the type of work on which he is engaged. I have found it a good policy, with all hand tools, to buy them as the need arises, rather than purchase several at the same time on the off-chance that they will come in handy in the future. This policy spreads expenditure evenly and ensures that only essential tools are purchased.

FILE LENGTH. The length of a file is measured from the shoulder above the tang to the point. Needle files are the exception to this rule, the total length being measured.

FILE CUT. A file may have a single or double cut. In single cut files, the teeth are parallel to one another and at an angle to the centerline of the file. The double cut file has a second cut over the first. This produces small pyramid-shaped teeth that have more cutting edges. It is the double cut file which is in common use. See Fig. 5.1.

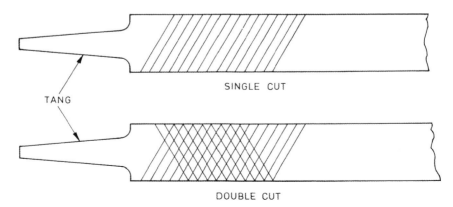

TANG

SINGLE CUT

DOUBLE CUT

Fig. 5.1 *Single and double cut files.*

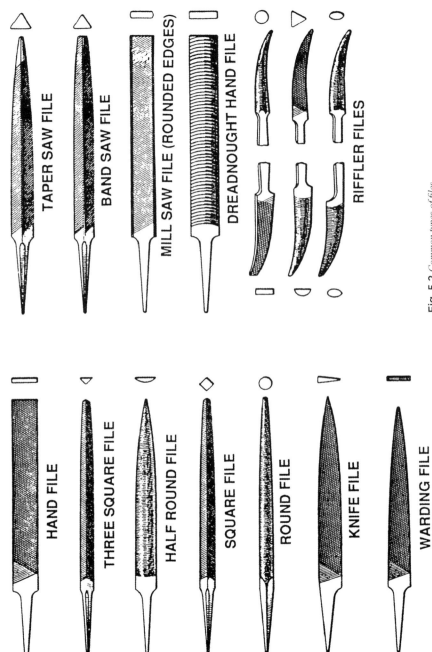

Fig. 5.2 *Common types of files.*

TAPER SAW FILE

BAND SAW FILE

MILL SAW FILE (ROUNDED EDGES)

DREADNOUGHT HAND FILE

RIFFLER FILES

HAND FILE

THREE SQUARE FILE

HALF ROUND FILE

SQUARE FILE

ROUND FILE

KNIFE FILE

WARDING FILE

Files may be cut with teeth of the following grades: rough, bastard, second cut, smooth, and dead smooth. Those in general use and likely to be found in the model engineer's workshop are the bastard, for the heavy removal of material but leaving a fairly rough finish; the second cut, a general purpose file for light removal giving a fair finish; and the smooth for fine finishing work.

Two rather special cuts should be mentioned, the dreadnought for the rapid removal of soft metal such as aluminum, and the rasp, which will deal, in a rough and ready sort of way, with wood.

FILE SECTIONS. There are several different file sections available to cope with different kinds of work. Fig. 5.2 shows the shape and section of a variety of files.

The **HAND FILE** is parallel throughout its length, viewed from the cutting face, but its thickness tapers toward the end. Both faces are double cut and one edge is single cut. The other edge is left smooth and forms a "safe" edge. This allows cuts to be made into corners without damaging the side against which the safe edge is in contact.

The **FLAT FILE** is tapered in both width and thickness and is double cut on both faces and single cut on its edges. It is a very useful general purpose tool.

The **HALF ROUND** is another useful shape, but misnamed; its section is not half round but only a segment of a circle.

SQUARE FILES are double cut on all four faces and are tapered for the first third of their length. They are useful to clear small corners and slots and to form square holes.

THREE SQUARE, or triangular files, are the specialty of the "saw doctor" but are also useful for shaping holes with less than right-angle corners, and for producing really sharp corners.

The **ROUND FILE**, colloquially known as the "rat tail," for obvious reasons, is tapered for the first third of its length. It is used for enlarging holes and comes in a variety of sizes. Buy one to suit your particular needs; I find one with a diameter of about ¼ in. (6mm) to be the most useful size.

WARD FILES are similar in shape and cut to the flat files but are much smaller. They are of uniform thickness throughout their length. Originally intended for use by locksmiths for producing wards in keys and locks, they are useful for dealing with the small components commonly found in the model engineer's workshop.

The **KNIFE FILE** is wedge shaped and not in general use but is useful on occasions in entering and enlarging slots where standard files are unsuitable.

NEEDLE FILES are very small files of various shapes and sections which are used for very fine and delicate work. The tang is formed into a thin cylindrical shape, and the pitch of the teeth range from 40 to 200 teeth per inch.

RIFFLER FILES are specially shaped to meet special requirements. They are, for example, used when tuning internal combustion engines to smooth the exhaust and inlet ports.

USE OF FILES

Files must never be used unless the tang is protected by a handle. The handle must be firmly attached to the tang and be a comfortable fit in the hand. Failure to fit the handle properly, or using a file without a handle, can lead to the tang being forced into the palm of the hand, causing a nasty injury. To fit the handle, the file should be held upright on a wooden bench or block of wood, tang uppermost, and the handle tapped firmly in position. The tang should extend well into the handle. The file itself must not be hammered. Files are dead hard and, in consequence, rather brittle. Any hammer blows are likely to cause chips to fly off the file, causing injury to anyone nearby and damage to the face of the hammer.

The beginner should learn the art of filing by practicing on odd pieces of scrap metal, mild steel, or cast iron, about 2 in. square in section being most suitable. A partly worn flat file should be used at first, as a new file is too sharp at this stage and in any case it is better reserved for use on brass or bronze before being taken into use on iron or steel.

The correct stance at the bench is just as important when filing as it is for sportsmen when playing golf or cricket. The handle of the file should be held comfortably in the right hand, and the left hand should grip the extremity of the blade with the fingers pointing downwards. The left foot should be placed well forward and the right foot turned slightly outwards. The body should be well balanced and one should feel comfortable. If you are left-handed, reverse the above instructions.

The file should be placed firmly on the work and moved forward with a firm, steady stroke. Every endeavor must be made to keep the movement in a horizontal plane, keeping even pressure on both ends of the file. Beginners nearly always rock the file so that the surface produced becomes curved instead of flat.

If the upper arm is considered as a lever, pivoted at one end at the shoulder joint and the other end at the elbow, it will be seen in Fig. 5.3 that if the shoulder is kept rigid, the elbow moves in an arc. Unless a positive effort is made to correct the movement of the file, it will rock, giving rise to a convex surface on the workpiece, which must be avoided.

A full stroke should be made with the file, in fact the stroke should only finish as the handle approaches the work. The file should then be drawn back, the pressure having been released. The file only cuts one way, the teeth being formed to cut on the forward stroke only.

The jaws of the vise in which the work is held should be level with the user's elbow. This is helpful in keeping the movement of the file horizontal. The work should be held securely in the vise. If the workpiece has a finished surface, or is soft, the clamps described in Chapter 1 should be used to prevent the serrations in the jaws of the vise causing damage.

Choose the right file for the job, use as big a file as is practicable, and do not nibble at the job with a small file. Use

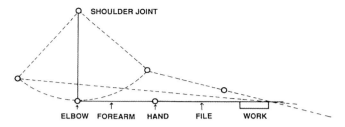

SHOULDER JOINT

ELBOW   FOREARM   HAND   FILE   WORK

Fig. 5.3 *Positive effort needed to prevent file rocking.*

a bastard file if a lot of metal has to be removed, then a second cut one. A smooth file may be used, as the work is brought to size, if a fine finish is required.

Files are commonly used at too fast a speed. A file is a cutting tool and, being made of carbon, not high speed steel, the correct cutting speed is quite low. It is difficult to give a hard and fast rule for the stroke rate for filing as there are many variable factors—the material being worked, the type of file being used, and the strength and experience of the worker all have to be considered. An average speed of around sixty to seventy strokes per minute is about right.

Keep checking the work with a straight edge and/or a square. Do this early and often, so that if an error is creeping in, it can be detected while there is enough metal left for a correction to be made.

By changing the direction in which the file is working, it is possible to check how the metal is removed by watching the marks made by the file. Fig. 5.4 shows a square piece of metal. If the file is used on the square from F to B, a set of file marks will appear at right angles to G—E. If the file is now used from E to A, it will be

found that the original marks will not be completely obliterated and it can be seen where the low spots are and the necessary corrections can be made. The file can then be used from G to C, where a fresh set of marks will appear. By constantly changing direction of the strokes of the file in this way, and by frequent checking of the work with a square and/or straight edge, accurate work can be produced. Slow, steady strokes with an even pressure will remove the metal at a good rate.

Use new files for such metals as brass and bronze, and when they have become dulled, they can be used effectively on cast iron and steel. Very old files should be used on castings where there may still be traces

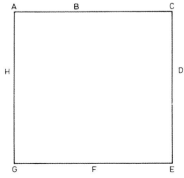

Fig. 5.4

Fig. 5.5 *Files hung on nails on a rack after holes have been drilled in their handles.*

surface of the work on which it is being used. If the offending metal cannot be removed by the file card in the ordinary way, it can be scraped out with a piece of pointed metal, being careful not to damage the teeth. Problems with pinning may be alleviated by rubbing chalk into the file, but this should not be done when filing cast iron or brass as it causes the metal to glaze under the file.

File marks may be removed by **DRAW FILING**. For this purpose a fine cut file is used crossway on the work, as is shown in Fig. 5.6. The file is grasped by the handle and tip and propelled forwards and backwards. The teeth of the file do not cut so harshly when it is used in this way.

Although filing in the lathe is not strictly benchwork, it must be considered. This practice does not find favor in some quarters, it being held that any finish required should be obtained by the lathe tools. However, the makers of Stubs files make particular mention in their user's handbook of filing in the lathe, and undoubtedly, in some circumstances, it is justified. They give the following advice:

"When work to be filed is revolving in the lathe, the file should be used with a stroking action, allowing it to glide slightly along the work. This will help to avoid making ridges and will help keep the file clear of chips. Because of their sharpness, new files are best avoided for lathework where a fine finish is required. Lathework should not be touched by hand, as oil and moisture can coat the surface and it is then difficult for the file to hold."

of the molding sand, which will spoil a new file. Avoid using files on sharp edges.

Files are expensive and pay for looking after. If they are thrown in a drawer with other tools, there is a danger that their teeth will be damaged. Hang the files on a rack; holes drilled in the handles allow them to be hung on nails on a rack, as is shown in Fig. 5.5.

Files should be regularly cleaned with a file card. This is a brush composed of stiff wire bristles woven onto a strong canvas. The canvas is nailed to a suitable wooden block, which forms a handle. The file blade should be brushed crossways with the card. This should not be done too vigorously so that the hard bristles blunt the file teeth. "Pinning" occurs when a piece or pieces of metal lodge between the teeth of a file, causing scratch marks to appear on the

Fig. 5.6 *Draw filing.*

There is some danger in filing in the lathe. Particular care must be taken to see there is no loose clothing likely to catch in the revolving chuck or work. Sleeves must be rolled up or tightly fastened at the cuff.

In order to produce a fine finish on work that has been filed, various grades of emery cloth are used. Emery cloth is obtainable from coarse grades right down to very smooth ones, and any of them is best used with an old file or a piece of wood as backing. If an especially fine finish is required, a thin oil on the emery cloth will assist. Worn-out coarse emery cloth should not be discarded but can be used as a fine grade.

Care must be taken when using emery cloth to ensure that the accurate work produced by the file is not ruined. Sharp corners can be rounded off by the careless use of emery paper.

Copper and its alloys can be given a high degree of polish when smooth by the use of a metal polish, such as Brasso.

Chapter 6

Hammers, Chisels, and Punches

HAMMER

The term "He is a hammer and chisel merchant" is sometimes used in a derisory way to infer a workman is prone to using brute force when something does not quite fit. Yet to use a hammer and chisel properly is a highly developed skill, only achieved after long practice.

Hammers come in a variety of shapes and weights. Engineer's hammers are known by the shape of the end opposite the striking face, the **PEIN**. The **BALL PEIN**, illustrated in Fig. 6.1b, is the most common type, the ball-shaped end being used mostly for riveting over the ends of rivets and pins. Cross and straight peins are useful for striking blows in awkward places.

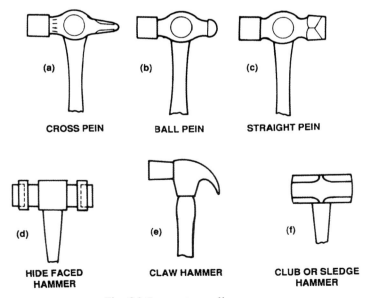

Fig. **6.1** *Common types of hammers.*

The **HIDE FACED** hammer, shown in Fig. 6.1d, has a hollow cylindrical steel center into which is pressed leather, or some similar material. In cases where blows are required on finished or semi-finished surfaces, the use of this type of hammer avoids damage. A plastic material is now often used for the striking face.

The **CLUB HAMMER,** shown in Fig. 6.1f, and the **CLAW HAMMER,** shown in Fig.6.1e, are not really engineer's tools. The claw hammer is a carpenter's tool. Instead of the pein described in the other types, it has a claw with a slot for drawing out nails. The club hammer is a mason's tool, but a larger hammer of this type, with a long handle, is known as the sledgehammer and is used by blacksmiths and on heavy engineering work. In use, when blows are struck, the shaft is held in both hands.

Hammer shafts are generally made of hickory. A slot is cut in the thin end of the shaft and this end is forced tightly into the hole in the hammer head. A steel wedge is then driven into the slot in the shaft, expanding it, so that it is a tight fit in the hammerhead. It is important that this tight fit of the shaft on the hammerhead is maintained. A head that is loose is likely to fly off when the hammer is in use, causing injury or damage to persons or things in the immediate vicinity.

Engineers' hammerheads vary in weight from 0.125kg (about 4 ounces) to 1.5kg (roughly 3½ lbs). Sledgehammers are made up to 14 lbs. The lighter hammers are used for the more delicate jobs but it

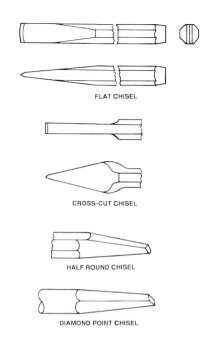

FLAT CHISEL

CROSS-CUT CHISEL

HALF ROUND CHISEL

DIAMOND POINT CHISEL

Fig. 6.2 *Types of chisels.*

is a mistake to use one that is too light. As experience is gained, it becomes an easy matter to pick the correct weight for the particular job in hand.

CHISELS

The cold chisel is a valuable tool for all sorts of reductions, alterations, and fitting of parts. Although now generally replaced by improved machine tools, they still have a place in every tool kit. They are called "cold" chisels because they are used for work on metals in their cold state. This is to distinguish them from tools which are used when working metals in a hot condition, such as in blacksmith's work.

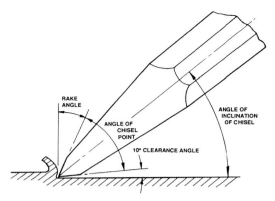

Fig. 6.3 *Cutting with a chisel.*

While the scraper, discussed in the next chapter, is used for the removal of small amounts of metal to make a precision fit, chiseling is the quickest way to remove metal by hand, but accuracy is not very high and surface finish is poor. It is basically a primary treatment before finishing off work with a file.

The various types of chisels are shown in Fig. 6.2. The **FLAT CHISEL**, the most common type, is used for surfacing and cutting off, the **CROSS-CUT** for roughing and grooving, and the **DIAMOND** for similar purposes. The **HALF ROUND** is particularly useful for cutting oilway grooves.

The action of cutting when using a chisel is shown in Fig. 6.3, the angles of rake and clearance being clearly shown. These angles depend on the angle of inclination of the chisel; if the angle is increased, then the clearance angle is also increased but the rake angle becomes less. For general use, a chisel point is ground to an angle of 60°. A clearance angle of

10° is suitable for most work, which means that the angle at which the chisel must be held must be 30° (half the chisel point angle) plus 10°, the clearance angle, which together equals 40°.

Soft metals, such as copper and aluminium, require a sharper chisel point than the harder ones. For example, an angle of 30° is suggested for aluminium, while 60° is suitable for cast iron and mild steel.

Chisels are made of tool steel, usually of octagonal cross section, but there is no reason why model engineers should not make them in the smaller sizes they require from the more readily obtainable round carbon steel. They must, of course, be hardened and tempered in the usual way.

After prolonged use, the head of the chisel becomes mushroomed in shape with jagged edges projecting all around the striking face (see Fig. 6.4). When reaching this state, the chisel is dangerous to use, as small pieces may fly off when it is struck by a hammer and an injury to the

Fig. 6.4 *Chisel with dangerous mushroom head.*

face or eye may result. In my own area, an unfortunate soul in a scrap yard lost the sight of an eye in this way. There is also the risk of the hand being injured by the sharp protruding edges. The jagged edges should be ground away on the emery wheel so that the head is restored to the shape shown in Fig. 6.2.

When first attempting chipping work, the beginner often misses the chisel and hits his hand. While in pain, and to avoid a repetition, he now watches the head of the chisel instead of the face of the work that is being chipped. This leads to further trouble as it is essential that the eye is directed at the actual formation of the chip.

The hammer shaft must not be gripped too tightly, but held comfortably and the blows struck with a wrist action. A tight grip is not needed on the chisel; small chisels can be held between the thumb and two fingers but chisels of ordinary size have to be fully grasped. All hand operations are difficult to describe, so if possible, watch an experienced person and copy his actions. In any cases "have a go," perhaps at first on scraps of mild steel that you do not mind spoiling.

Exercise enough control of the chisel to keep it at the right angle, which has to be modified during progress across a face, according to how the cutting edge of the chisel is penetrating the work. Lower the angle if too much digging in has developed

and raise it slightly if penetration is not deep enough.

The edges of metals, particularly the crystalline kind such as cast iron, are liable to break away as the chisel reaches them. It is wise, therefore, to stop the cut before the edge is reached, and then reverse the direction of the cut, that is to chisel inwards.

In the past, when a large area had to be reduced, grooves were cut across the face using a cross-cut chisel at a distance from one another slightly less than the width of the available flat chisel (see Fig. 6.5). The flat chisel was then used to bring the whole surface down to the level of the bottom of the grooves. With the machine tools now available, I doubt if this is now ever done but the method is worth bearing in mind for use in an emergency when no other means of completing a job is at hand.

A flat chisel is a very useful tool for cutting sheet metal, particularly if the job is small enough to be held in the vise. The

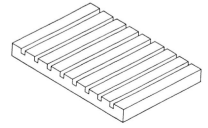

Fig. 6.5 *Grooves cut with cross-cut chisel ready for finishing with flat chisel.*

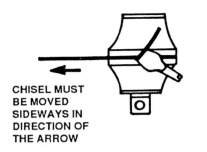

**CHISEL MUST
BE MOVED
SIDEWAYS IN
DIRECTION OF
THE ARROW**

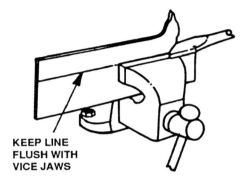

**KEEP LINE
FLUSH WITH
VICE JAWS**

Fig. 6.6 *Cutting sheet metal in the vise.*

scribed line, indicating where the cut has to be made, is placed level with the vise jaws. The chisel is then applied, using the vise as a guide, so that a shearing action occurs. With care, a clean cut can be made (see Fig. 6.6).

For anyone restoring old engines, the chisel is a very valuable tool for such tasks as cutting off rivet heads and splitting seized nuts. I think it got its bad name from the practice of some wood-carvers using it to slacken a nut when a spanner of the correct size is unavailable. The condition of the nut after being mutilated in this way is a very sad sight. Properly used, the chisel is a craftsman's tool not to be despised.

PUNCHES

The **CENTER PUNCH** is an essential tool in the fitter's toolkit. It is made of tool steel and is hardened and tempered in the same way as a chisel. Its point is ground to a fine conical shape (see Fig 6.7). It is used for marking the center point of holes to be drilled. As well as locating the position

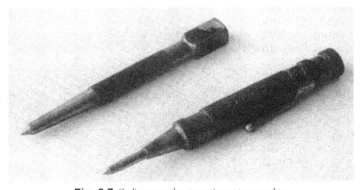

Fig. 6.7 *Ordinary and automatic center punches.*

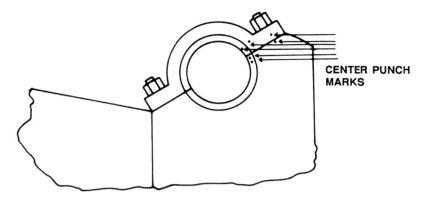

CENTER PUNCH
MARKS

Fig. 6.8 *Main bearing marked with center punch to ensure correct assembly.*

of the hole, it also prevents the drill from wandering from its position during the starting process. It is also used to provide a point in which to place one leg of the dividers when a circle has to be scribed.

To ensure parts being dismantled are reassembled in exactly the same position, it is good practice to mark their position with center punch marks. Fig. 6.8 shows how two bearing brasses are marked so that they are always assembled correctly.

In use, the center punch is held between the thumb and the first two fingers of the left hand in a vertical position with the point on the exact spot where the mark is to be made. It is then given a light blow with a hammer held in the right hand.

A **DOT** or **PRICK PUNCH** is similar to a center punch, but is lighter with a finer point. It is used to mark the position of scribed lines so that the line may be restored if it becomes obliterated. This subject will be dealt with further in the chapter on marking out.

The **AUTOMATIC CENTER PUNCH** has an internal spring mechanism.

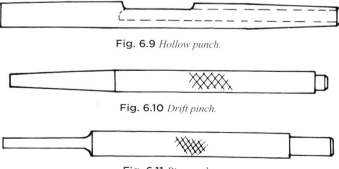

Fig. 6.9 *Hollow punch.*

Fig. 6.10 *Drift pinch.*

Fig. 6.11 *Pin punch.*

The point of a punch of this type is placed in position on the workpiece and pressure is applied to the top of the punch. This compresses the spring until the punch reaches a certain position where a catch is released. The energy from the spring is transmitted to the point of the punch and a light center punch mark is made. Generally, the punch has an adjustment that regulates the force of the blow so that center punch marks of varying depths can be made. See Fig. 6.7.

The **PIN PUNCH** (see Fig. 6.11) has a parallel round shank end instead of the tapered end on the center punch. It is used for driving out pins, rivets, or bolts from components. The shank at the end varies in size and, when used, a size a little smaller than the piece to be removed should be selected.

DRIFT PUNCHES are similar to pin punches, except that instead of a parallel shank at the end, the end tapers. It is a general purpose tool with a variety of uses. It is particularly useful for imparting a blow in a confined space that cannot be reached directly by a hammer. Driving a ball race on to a shaft or into a housing is a typical example of the type of job for which the drift punch is used (Fig. 6.10).

When cutting gaskets, it is generally necessary to produce a number of holes in the gasket material to match those in the joint faces. **HOLLOW PUNCHES** in various sizes are used for this purpose, (see Fig. 6.9).

The punch has a hole in its end that extends some way into the body of the punch. A slot is machined in the side of the punch to meet the hole. The gasket material is placed on some relatively soft material, such as lead or a piece of hardwood, and the punch is placed in position where the hole is required and then struck with a hammer. A hole is made in the gasket material and the piece of material that has been removed travels up the hole in the punch, eventually being discharged through the slot. The gasket material must not be placed on steel, or similar hard metal, when gaskets are being made, as the punch will be blunted.

Chapter 7

Scrapers and Scraping

Apparently, flat surfaces are almost certain to have small irregularities, even though they have been carefully filed or machined. There are high spots and valleys, and if the high spots are left between moving, mating surfaces, they will rapidly wear down, leaving excessive clearance between the parts.

Scrapers are used for removing these high spots, and in the hands of a skilled fitter, small shavings of metal can be removed with the scraper as and where required. Internal cylindrical surfaces, such as engine main and big end bearings, can also be scraped so that they are a very good fit on their journals.

Scrapers are made in three shapes, flat for dealing with flat surfaces, half round for producing good mating surfaces between a shaft and its bearing, and a three-corner one. The last can be used in the same way as the half round one and it is also useful for de-burring holes.

Because of the high quality steel from which files are made, old, worn-out flat and half round files make very good scrapers. The file should be softened by heating in a fire to a red heat, and then allowed to cool slowly in the ashes. It can then be forged and filed to the desired shape, and finally re-hardened and tempered. This hardening and tempering process will be discussed later.

Since the cutting force on a scraper is comparatively light and without shock, its edge need only be given a slight degree of temper. This extra hardness is a distinct advantage because the cutting edges of scrapers tend to become blunt very quickly. An illustration of a flat scraper appears in Fig. 7.1.

If care is taken not to draw the temper of a flat file by letting it get too hot when shaping it on the emery wheel, it is possible to make a flat scraper without softening and re-hardening it. The teeth on the file are ground away and the end slightly radiused to prevent it digging in when in use.

Once the cutting edge has been produced by grinding, it must be sharpened on an oilstone. It is sharpened by holding it close to its cutting edge and adopting a rocking motion, and by stoning the sides of

25MM

Fig. 7.1 *Flat scraper.*

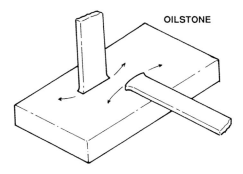

OILSTONE

Fig. 7.2 *Sharpening a flat scraper.*

the cutting edge until they are dead sharp (see Fig. 7.2). The cutting edge must be kept very sharp so it is necessary to be constantly touching up the scraper on the oilstone as the work progresses.

The scraper only removes slight irregularities; it is not intended to remove much metal. Assuming the surface is reasonably flat, another perfectly flat surface is required with which to compare the surface being scraped. For this purpose a surface plate is needed. Traditionally, it is rectangular in shape and made of good-quality, close-grained cast iron. It

has several stiffening webs on its base to prevent any tendency for it to "sag" (see Fig. 7.3). However, granite surface plates, with their greater wear resistance, are becoming more popular and may be the only choice when purchasing new.

Surface plates are expensive and are not common in the model engineer's workshop. A sheet of thick plate glass is a very good substitute, and an off-cut can probably be obtained from a builder's merchant with a glazier's department. I was fortunate in obtaining a piece with ground edges, but if this is not available, it is not a difficult job to provide a baseboard for the glass to sit on, and to surround the glass with molding not quite so deep as the glass is thick.

The surface plate, or glass, is thoroughly cleaned and then evenly smeared with engineer's blue. A tube of Prussian blue artist's oil paint will do if the proper "blue" is not available. The surface to be scraped is then placed on the surface plate and rubbed about slightly. All the high

Fig. 7.3 *Surface plate.*

40

spots will now appear coated blue. These high spots are reduced in height by careful scraping as shown in Fig. 7.4.

The handle of the scraper is held in the right hand and the left hand presses on the blade. Short strokes are made, ½ in. (12mm) being about the correct length. The scraper is kept in contact with the surface of the work on the return stroke but no pressure is applied to it. Having gone over the work with strokes in one direction, the next set of strokes should be made at right angles to the first.

The work is cleaned off with a rag, the blue again evenly smeared on the surface of the surface plate, and the work rubbed gently on the surface plate once more. The high spots now appearing are treated in the same way as before, and the whole process repeated until the surface of the work is covered by small areas of contact fairly close to one another.

It is a mistake to apply the blue too thickly to the glass or the surface plate as a false reading may be obtained. Much

patience and practice are necessary to produce accurate surfaces by scraping, but if the tool is kept very sharp and gentle pressure is used, good results should be obtained. The frost effect, so much admired on the beds and slides of high-quality machine tools, can then be attempted.

If accuracy is not important and only appearance is to be considered a frosted effect can be obtained by cheating a little. Glue a ¼ in. diameter circle of emery cloth on the end of a piece of ¼ in. wooden dowel, and place in the chuck of the bench drill. Oil the work to be treated and, with the drill running at high speed, bring the emery into light contact with the work. Move the work along and bring the emery into contact with the work again. Repeat the process until the whole of the area is covered with a series of shiny circles. The purist may hold up his hands in horror but the treatment does give the work a nice appearance, and for some reason, which I cannot explain, makes ferrous metals corrosion resistant. Naturally, discretion

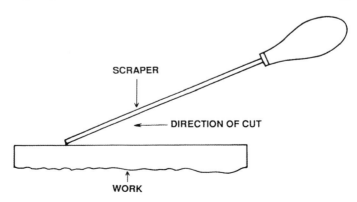

Fig. 7.4 *Scraping a flat surface.*

41

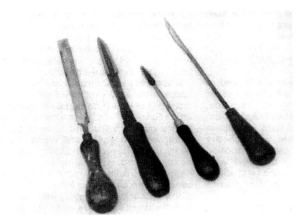

Fig. 7.5 *Selection of scrapers. Note all are fitted with handles.*

must be used with this practice and it is not recommended where accuracy is important.

The half round scraper can be made from a half round file. Scrapers are available commercially but there is much greater satisfaction when using one you have made yourself than one you have bought from a shop. The file should be softened, as described, and then forged and filed to shape. The center of the working face is hollowed so that there is less metal in contact with the oilstone when the scraper is being sharpened. Get as fine a finish as possible, finishing off with fine emery paper. It must then be hardened and tempered and sharpened on the oilstone.

When I was an apprentice, I made a half round scraper from the outer ring of a ball race. The ball race was dismantled and the outer ring softened by heating in the forge and allowing to cool slowly. It was then cut, straightened out, and forged roughly to shape. The groove in which the balls used to run provided the correct hollow ground effect. It was filed to the correct shape and the firm's blacksmith offered to harden and temper it for me. He was a craftsman of the old school and got the degree of hardness just right. I have used this tool for many years and it is still as good as new. I particularly remember during the war years, when I helped to keep an essential fleet of motor cars running when new cars and spares for the old ones were unobtainable, using the scraper to scrape out the ridges at the top of the cylinder bores. The piston ring grooves were machined to fit new rings, and these would have fouled the ridges as the pistons reached the top of their stroke if the ridges had not been removed.

The half round scraper is used for finishing the surface of bearings so that they are a good fit on their shaft. The shaft is smeared with a thin layer of "blue," placed in position on the half bearing and rotated a few times and then

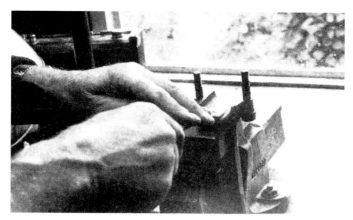

Fig. 7.6 *White metal big end bearing being scraped.*

taken out. The high spots on the bearing will be marked with blue and these marks are removed by the scraper. This process is repeated until the whole area of the half bearing shows contact with the shaft. When the bottom half bearing is finished, the other half is placed in position and treated in the same way, only in this case the two halves are bolted together. As well as getting a good surface fit, the bearing must not be too tight or slack on its shaft. Shims (thin pieces of metal, usually brass) are fitted between the half bearings if the bearing is too tight, and the half bearing carefully filed if it is too slack.

A selection of scrapers is shown in Fig. 7.5, and Fig. 7.6 shows a white metal big end bearing being scraped. The right hand keeps the tool steady and exercises a certain amount of control and movement over the tool, while the fingers of the left hand are placed on top of the scraper close to the work. The tool is drawn steadily across the work with slight pressure applied to enable the cutting edge to make the cut. Successive cuts should be made in opposite directions so that any high spots left from the previous cut are removed.

Chapter 8

Measuring

Accurate measurement is the basis of good engineering practice. In the building trades craftsmen are content to use a wooden ruler, but the engineer always uses a precision engraved steel rule, and even this is used only for the least important measurements. Vernier calipers or micrometers are used when greater accuracy is required.

It could be argued that we should all be using the metric system of measurement but many workshops are still committed to the imperial system of feet and inches and, as far as possible, I shall be considering both systems.

A 6-inch and a 12-inch steel ruler with both metric and imperial markings will cover most needs. One graduated in ⅛, 1/16, 1/32, and 1/64 in. on one side and millimeters

and ½ millimeters on the other is ideal. A pocket steel tape, about 6 feet long, is useful for some classes of work.

The degree of accuracy to which work may be produced when only a ruler is used depends on the quality of the ruler and the skill of its user. It is extremely difficult to guarantee an accuracy much closer than within 1/64 in. or 0.3mm.

It is important to own a good ruler and to get used to using it. It must be looked after: the end particularly should be preserved from wear, as it generally forms the basis for one end of the dimension. Rulers should be oiled to prevent rusting when not in use. Fig. 8.1 shows a selection of rulers.

The **TRY SQUARE** is the most common tool for testing squareness and Fig. 8.2 shows it in use. Try squares are precision instruments and must be treated with great care if they are to retain their initial accuracy. They must not be knocked or dropped and should be kept away from bench tools to avoid burrs occurring on the blade edge. They must be checked at regular intervals for squareness. This is done by the simple method illustrated in Fig. 8.3. Two lines are scribed in the manner shown, and if the lines are not one

Fig. 8.1 *A selection of steel rules.*

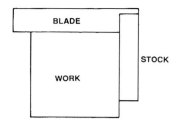

Fig. 8.2 *Try square in use.*

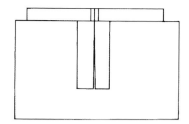

Fig. 8.3 *If scribed lines do not coincide when drawn with square in opposite positions the square is out of true.*

over the other, the square is faulty. This check is quite accurate, as the error shown is double the error present in the square.

When two surfaces are at any angle other than 90°, the angle between them can only be measured with some form of **PROTRACTOR**. One with a graduated head is shown in Fig. 8.4. A more sophisticated type has a vernier attachment whereby very accurate readings may be obtained.

The **BEVEL**, shown in Fig. 8.5a, is a useful tool that is very easily made. In fact, it forms a useful exercise in drilling and filing, and a drawing for its construction is also shown in Fig. 8.5b. It cannot be set to a desired angle without some other aid and this limits its scope.

The **COMBINATION SQUARE** (Fig. 8.6) manufactured by Moore & Wright, a subsidiary of Neill Tools, is a versatile instrument. It consists of a hardened steel rule with imperial graduations in 64ths of an inch and metric ones in ½mm. The square head has a fixed angle of 15°, 30°, 45°, 60°, and 90°. The center head has a fixed angle of 90° for finding the center of round or square bars. The protractor has a full 360° scale graduated 0° to 180° to 0°.

It is satin chrome finished for easy reading. Fig. 8.7a shows some of its uses, while Fig. 8.7b shows it being used to measure an acute angle.

CALIPERS AND DIVIDERS

Calipers are used for measuring distances between or over surfaces, or for comparing distances or sizes with standards such as those on graduated rulers. Their shape varies according to whether they are for measuring internally or externally. Simple firm leg calipers are shown in Fig. 8.8 and

Fig. 8.4 *Protractor with graduated head.*

45

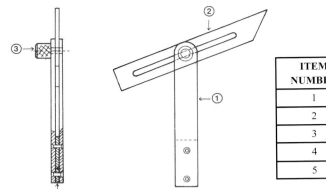

ITEM NUMBER	DESCRIPTION
1	BODY
2	BLADE
3	SCREW
4	DISTANCE PIECE
5	RIVETS—2 OFF

Fig. 8.5a *Bevel.*

spring-operated calipers with a screw adjustment appear in Fig. 8.9.

With care and skill, very accurate results can be obtained with calipers; in fact, the old millwrights used them exclusively for measuring large diameters and obtained very accurate results. A very light hold at the caliper joint is essential. The action of sliding over the object to be measured must be a very delicate one so

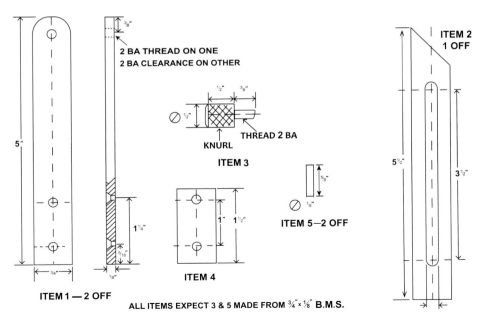

Fig. 8.5b *Bevel details.*

Fig. 8.6 *Combination square.*

MITER 45°

DEPTH GAUGE

ABUTMENT CHECKS

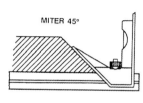

Fig. 8.7a *Various uses of the combination square.*

CENTER LINE OF DISC

PROTRACTOR ACCESSIBILITY

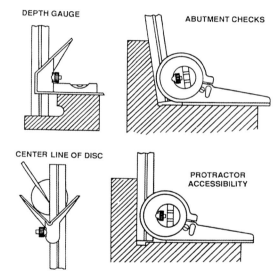

Fig. 8.7b (below). *Combination square used to measure an acute angle.*

ACUTE ANGLE

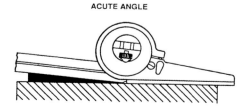

that the sense of touch notes the faintest difference in pressure at the caliper points. Failure to operate in this gentle manner will cause the points of the caliper to be forced over the work, springing the legs apart and giving a false reading.

It is more difficult to measure bores accurately because of the difficulty in setting the points of the calipers across the

47

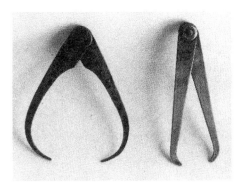

Fig. 8.8 *Firm leg calipers.*

absolute center of the bore. If care is taken in pushing the caliper into the bore and wriggling it up and down and sideways, the highest point can be felt.

Calipers can be set from a sample of work, from a ruler or from a micrometer. Probably the most accurate setting can be obtained from a sample piece of work. In setting from a ruler it is necessary to be very careful, first, not to use an old ruler that has become worn at the end, and second, to set the caliper point so as to exactly split the line of graduation on the ruler.

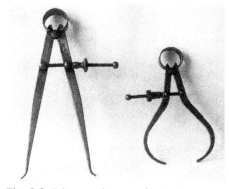

Fig. 8.9 *Calipers with screw adjustment.*

Quite accurate results can be obtained by setting calipers using a micrometer. This is often useful when it is necessary to measure the inside diameter of a bush or cylinder. The inside calipers are used to measure the bore, and then the calipers are measured with the micrometer. If the same care is taken as was recommended in the previous paragraphs, very accurate results can be obtained.

DIVIDERS (Fig. 8.10) are used for scribing circles and marking off lengths. **HERMAPHRODITE CALIPERS**

Fig. 8.10 *A pair of dividers and a hermaphrodite caliper, better known as "odd legs" or "Jennies."*

are half calipers and half dividers. More generally known as **ODD LEGS** or **JENNIES**, their use and the use of dividers will be dealt with in the chapter on Marking Out.

MICROMETERS. Where very accurate measurements are required, it is necessary to use a micrometer with which it is possible to make readings to 0.001 in. or 0.01mm. An illustration of a Moore

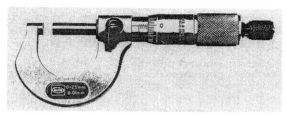

Fig. 8.11 *Moore & Wright metric micrometer.*

& Wright metric micrometer appears in Fig. 8.11. The inch type is of similar construction but the graduations, quite obviously, are different.

The micrometer consists of a semicircular frame having a cylindrical extension, the sleeve, at its right end and a hardened anvil at the other end. The bore of the sleeve is threaded and a spindle screws into the bore. The spindle carries a graduated thimble that turns at one with it.

Imperial

To read in one thousandths of an inch (0.001″).

First, note the whole number of major divisions (tenths of an inch or 0.1″ shown on the sleeve)	**Two × 0.1 = 0.2″**
Then note the number of minor divisions after the whole number (each minor division is equal to a quarter of a major division, i.e. twenty-five thousandths of an inch or 0.025″)	**Three × 0.025 = 0.075″**
Finally, read the line on the thimble coinciding with the datum line. This gives thousandths of an inch.	**Eleven × 0.001 = 0.011″**
	TOTAL = 0.286″

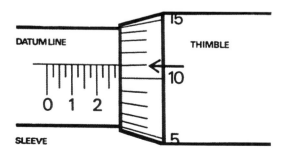

Fig. 8.12 *Reading the inch micrometer.*

Metric

Reading in hundredths of a millimeter (0.01mm).

First, note the whole number of mm divisions on the sleeve (major divisions—below datum line)	**Ten × 1.0 = 10.0mm**
Then observe whether there is a halfmm visible (minor divisions above datum line)	**One × 0.5 = 0.5mm**
Finally, read the line on the thimble coinciding with the datum line. This gives hundredths of a mm.	**Thirty-three × 0.01 = 0.33mm**
	TOTAL = 10.83mm

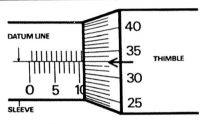

Fig. 8.13 *Reading the metric micrometer.*

READING THE "INCH" MICROMETER. The inch reading micrometer screw has 40 threads per inch, so that in one complete revolution it moves ¼₀ in. (0.025 in.). In ¹⁄₂₅th of a turn, it will move ¹⁄₂₅ of ¼₀ in., which is 0.001 in. The sleeve has marked on it major divisions, representing tenths of an inch—that is 0.100 in. Each major division is sub-divided into four minor divisions, representing 0.025 in. each.

The thimble is divided into twenty-five parts around its beveled circumference, and as one full turn is equal to one minor division on the sleeve (0.025 in.), then one division on the thimble will be 0.001 in. Fig. 8.12 shows how 0.286 in. reads on a micrometer.

READING THE METRIC MICROMETER. The screw on metric micrometers has a pitch of ½mm, so two revolutions of the thimble will move the spindle through 1mm. On the sleeve, the datum line is graduated with two sets of lines:, the one below the datum reading in millimeters and the set above reading in half millimeters. The thimble scale is marked in fifty equal divisions, figured in fives, so that each small division on the thimble represents ¹⁄₅₀ of ½mm, which equals ¹⁄₁₀₀mm, which is 0.01mm.

To read the micrometer, first note the whole number of millimeter divisions on the sleeve (major divisions), then observe if there is a half millimeter visible (minor divisions), and lastly, read the thimble for

Fig. 8.14 *Vernier caliper.*

hundredths (thimble divisions) i.e., the line on the thimble coinciding with the datum line. This procedure is plainly set out in Fig. 8.13.

An inch and a two inch micrometer or their metric equivalents will cover most needs, although larger sizes are available. The larger ones often have a frame into which various length anvils can be inserted to cater for the different lengths to be measured. The actual measuring mechanism is the same as it is on the smaller sizes.

THE VERNIER CALIPER, Fig. 8.14, is used for measuring work where the micrometer cannot be used, or for work outside a micrometer's capacity. Most verniers can measure inside, outside, and depth dimensions. It consists of a beam that has a fixed jaw at one end and also a sliding jaw. The beam is graduated with both imperial and metric scales.

READING THE INCH VERNIER.

The main scale on the vernier is graduated and numbered in inches, with each inch graduated and numbered in tenths (0.1 in.). Each tenth is divided into four, giving 0.1 in. divided by 4 = 0.025 in. On the sliding jaw, 0.6 in. is divided into 25 parts. Each of these has a length of 0.6 in. divided by 25 = 0.024 in. The differences in length between a small division on

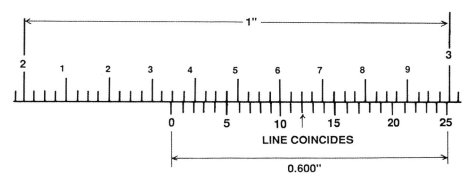

Fig. 8.15 *Reading 2" + 0.300" + 0.025" + 0.012" = 2.337".*

the fixed scale and the sliding scale is 0.025–0.024 = 0.001 in.

If the zero on the main scale and the zero on the sliding scale are level, and then the sliding scale is moved until the first small marks on both scales are level, the movement will have been 0.001 in. If the sliding jaw is moved until the 16th mark on the sliding scale is exactly level with a mark on the main scale, the sliding scale has moved 0.016 in. and so on.

To read a measurement, note the position of the zero line on the vernier scale in relation to the main scale. In Fig. 8.15, this is shown as 2.00 in. plus 0.300 in. plus 0.025 in. which equals 2.325 in. To this must be added the number of divisions from the zero line on the vernier scale to the line that is coincident with a line on the main scale. In this case, 12 divisions which equals 0.012 in. the total reading is, therefore:

• Main scale	• 2.325 in.
• Vernier scale	• 0.012 in.
Total	2.337 in.

READING THE METRIC VERNIER.
The main scale on the metric vernier is graduated in millimeters and numbered every ten divisions. The vernier scale is divided into fifty divisions over a distance of 49mm, each division equaling $^{49}\!/_{50}$th of a millimeter (0.98mm). The difference between a division on the main scale and the vernier scale is $^{1}\!/_{50}$th millimeter (0.02mm).

The metric vernier is read in a very similar way to the inch one already described. To read the measurement, note the main scale measurement immediately preceding the zero line on the vernier scale. To this must be added the decimal reading on the vernier scale. Note the line on the vernier scale that is coincident with a line on the main scale. Let us assume that this is the third line after the decimal number 0.7. Then we shall have to add 0.7 plus three divisions of 0.02mm = 0.76mm. This figure is then added to the main scale reading.

It cannot be emphasized too strongly that measuring and testing tools are precision instruments that must be treated with great care. They should not be left lying on the bench where they can become damaged by coming into contact with other tools or workpieces. If a protective case or box is provided, they should be returned there after being carefully cleaned. If the tool is likely to be out of use for some time, a thin coat of a non-corrosive oil should be applied to the measuring faces and bright spots.

Chapter 9

Marking Out

Except for very simple operations, such as filing the end of bar material square, it is necessary before starting work to scribe lines indicating the profile or outline of the finished article and the position of any holes. This process is known as "**MARKING OUT**."

A special cellulose lacquer, usually blue in color, is available for applying to any bright surface so that scribed lines show up clearly. When applied thinly, it dries very quickly. It must be kept in an airtight container to prevent evaporation; sometimes it is supplied in a properly designed dispenser bottle complete with brush. It can also be obtained in an aerosol container so that it can be sprayed on the work, but this is rather wasteful as much of the spray misses the target.

Rough castings and forgings are usually painted with a light distemper or emulsion paint thinned down with water. Recently, someone recommended typewriter correcting fluid. I know this dries very quickly, but I have not tried it out for marking-out purposes. I should think it is rather an expensive method compared with emulsion paint.

Although some of the tools used for marking out are used exclusively for this purpose, others are in general use in the workshop. The **SURFACE PLATE**, for example, was described in Chapter 7 when scraping a flat surface was discussed.

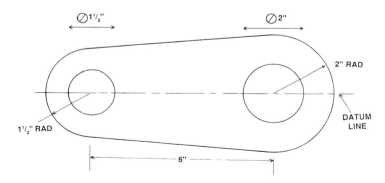

Fig. 9.1 *Marking-out exercise on link.*

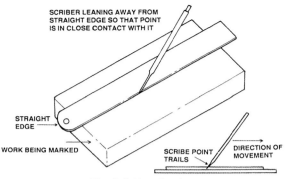

SCRIBER LEANING AWAY FROM
STRAIGHT EDGE SO THAT POINT
IS IN CLOSE CONTACT WITH IT

STRAIGHT
EDGE

WORK BEING MARKED

SCRIBE POINT
TRAILS

DIRECTION OF
MOVEMENT

Fig. 9.2 *Use of scriber.*

Dividers, the odd leg calipers, and the try square were all dealt with in Chapter 8 on "Measuring."

Before marking out, a datum line or surface must be established. A simple example of the use of a datum line is shown in Fig. 9.1. The link illustrated is to be made from ½ in. steel plate. The center line is the datum line in this instance and is scribed using a **SCRIBER**. The scriber must have a fine, sharp point kept in good condition on an oilstone; a grinding wheel should not be used for this purpose. Gramophone needles, as used on the old 78 RPM machines, make excellent scribers when held in a pin vise. Unfortunately, they are now difficult to obtain, but ready-made scribers are not very expensive. The correct method of using a scriber is shown in Fig. 9.2.

Getting back to marking out the link, after the centerline has been scribed, the centers of the two holes are marked out. Having made a light center-punch mark locating the center of one of the holes, the dividers are set to 6 inches, and the position of the second hole is found by placing one

leg of the dividers in the center-punch mark and scribing an arc on the centerline. A center-punch mark is then made where this arc meets the datum line. The outlines of the two holes are now scribed in, with the dividers, one with a radius of 1 in. and the other with a radius of ¾ in. The 2 in. radius at the end of the link with the larger hole is now scribed in followed by the 1½ in. radius at the other end. These two radii are then joined with tangential lines using a ruler to guide the scriber. The scribed lines can be made more permanent with light center-punch marks. The depth of the two center-punch marks in the centers of the holes should now be increased to allow the drill to make an accurate start.

To avoid error, each feature of the component should be marked out from the datum line or surface and not from one another. Take, for example, the series of holes shown in Fig. 9.3, which are required to be 25mm apart. Let us suppose that the dividers are set, in error, at 25.2mm. Then, if each hole is marked off from the one next to it, the error in the position of

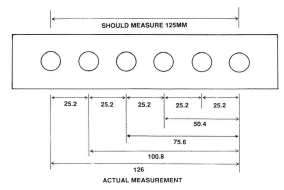

SHOULD MEASURE 125MM

25.2   25.2   25.2   25.2   25.2

50.4

75.6

100.8

126

ACTUAL MEASUREMENT

Fig. 9.3 *Cumulative error caused by not measuring from datum line.*

the last hole will be 1mm; i.e., it will be 126mm away from the first hole instead of the correct dimension of 125mm.

The link we marked out was only "flat" work, and when solid objects have to be dealt with, additional equipment is required. In the first place, a flat surface on which the work can stand must be provided and the **SURFACE PLATE** is used for this purpose. As stated earlier, a piece of plate glass makes an excellent substitute if a surface plate is unavailable.

For scribing lines at a given height above the base, the **SURFACE GAUGE**, Fig. 9.4, is used. The spindle can be adjusted to any height, and then sensitively adjusted to position by the knurled nut. The head carrying the scriber is so made that when the clamp nut is loosened the scriber can be freely moved to any position on the spindle.

The height of a scriber point on a surface gauge is set against a ruler held truly vertical to the surface plate by being pressed against an angle plate. Alternatively, a combination square can be used, the ruler being held vertical with the square head resting on the surface plate.

The **ANGLE PLATE**, Fig. 9.5, is used for supporting a surface at right angles to the surface of the surface plate. It is provided with holes or slots to accommodate bolts needed to secure articles to it.

Fig. 9.4 *Surface gauge.*

55

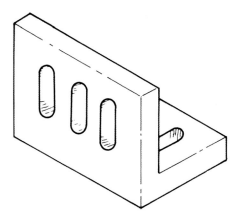

Fig. 9.5 *Angle plate.*

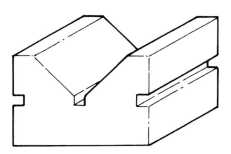

Fig. 9.6 *V-block.*

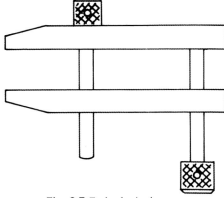

Fig. 9.7 *Toolmaker's clamp.*

The **VEE BLOCK**, Fig. 9.6, is used for supporting shafts and bushes. They usually come in pairs and a clamp is generally provided for holding work in the "V."

TOOLMAKER'S CLAMPS, Fig. 9.7, are essential items in a fitter's tool kit. They are used for clamping work to an angle plate when marking out and for many other purposes. They are very useful, for example, for clamping two pieces of work together when holes have to be drilled in alignment in both pieces.

FEELER GAUGES, Fig. 9.8, consist of a series of thin blades of different thicknesses, housed in a container in much the same way as the blades of a penknife. There are about ten blades in each tool and each blade has its thickness marked upon it. The blades on an imperial set are invariably marked in thousandths, their thickness varying from 0.0015 in. to 0.025 in. The blades on a metric set are marked in 100ths of a millimeter and range from 0.05mm to 0.8mm. They are used to gauge small distances between adjacent surfaces.

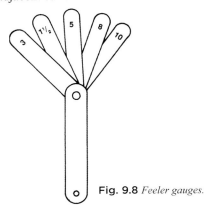

Fig. 9.8 *Feeler gauges.*

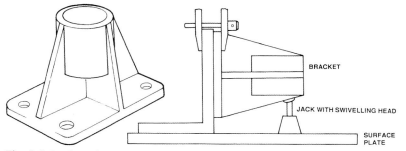

Fig. 9.9 *Bracket to be marked out.*

Fig. 9.10 *Bracket set up for marking out.*

When marking out on a surface plate, steps must be taken to ensure that the component to be marked out sits firmly on the surface plate without any tendency to rock. Sometimes it is necessary to file or machine the surface lightly to achieve this. In other cases, packing pieces or small jacks are used to keep the casting resting firmly in the required position.

A set-up for marking out the bracket shown in Fig. 9.9 appears in Fig. 9.10. The base of the casting is clamped to an angle plate with toolmaker's clamps, while a small jack with a swivel head supports the inclined surface. The small holes will be plugged with disks of wood to allow centers and lines to be marked, while a bar of wood or steel will span the large cored hole. Fig. 9.11, where a cylinder is being marked out, shows a strip fitted across the bore of the cylinder to provide a surface for the center-punch mark necessary for scribing the circle showing the outline of the cylinder bore. Horizontal lines are scribed with the surface gauge and vertical ones by using a set square and scriber or, in some cases, the odd leg calipers.

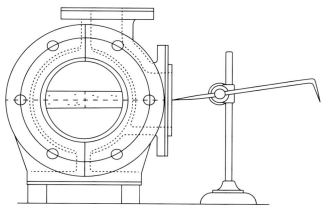

Fig. 9.11 *Cylinder being marked out. Strip across bore for dividers to mark out circle showing position of bore.*

Chapter 10

Drills and Reamers

Much of an engineer's time is spent drilling holes, and to obtain the best results—that is to produce a hole of the correct size, in exactly the right place, with a good finish—requires a certain amount of expertise.

Years ago, flat drills, of the type shown in Fig. 10.1, were used but, although they are still handy for special jobs and can be made in the workshop, they have serious drawbacks. The advent of the twist drill, of the type used in industry and by model engineers, was an important advancement.

Fig. 10.2 shows a Dormer drill and gives twist drill nomenclature. The drill illustrated is provided with a Morse

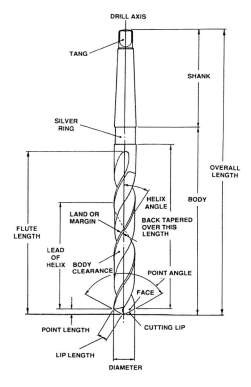

Fig. 10.1 *Flat drill.*

Fig. 10.2 *Dormer drill with Morse taper shank.*

taper but, of course, drills with straight shanks that can be held in a drill chuck are also made.

Most workshop drilling will be done in a pillar or bench drilling machine. Fig. 10.3 shows a machine of the latter type in use. When the drilling machine cannot accommodate the work or the work cannot be brought to the machine, a portable drill, either hand, breast, or electric, is used (Fig. 10.4). With these types, difficulty may be experienced in keeping the drill square with the work, and a small square is useful in checking that the drill is at right angles to the work. It is an advantage if this checking can be done by a helper, as it is difficult to operate the drill and see that it is square with the work at the same time.

Fig. 10.3 *Bench pillar drill.*

Standard twist drills are ground with an included angle of 118°, established as the most suitable for general purpose work. If the correct initial clearance is produced and increased gradually toward the center to produce a chisel edge angle of approximately 130°, the correct clearance will be achieved along the whole of the cutting lip. The two cutting lip lengths should be equal and at a similar angle to the drill axis to provide correct balance and concentricity (see Fig. 10.5).

Fig. 10.4 *Hand, breast, and electric drills.*

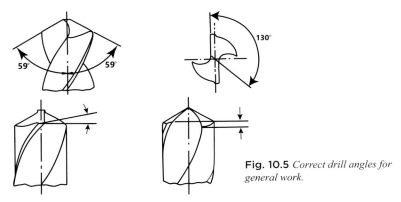

Fig. 10.5 *Correct drill angles for general work.*

It is recommended that the angle of drills be altered for drilling certain metals and for some special jobs, but it is not generally necessary unless a great deal of drilling of these metals is done. The standard 118° angle drill will cater for most jobs if care is taken.

To sharpen a drill by hand and to produce the correct angles is extremely difficult, but at the same time, accuracy is essential. For example, if the lips of the drill are of different lengths, an oversize hole will be produced (see Fig. 10.6). There is no doubt that a well-designed drill grinding device is a great asset to any workshop. I

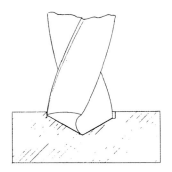

Fig. 10.6 *Drill lips of different lengths produce oversize holes.*

have found the Newmartek drill sharpener very satisfactory. It will be seen from Fig. 10.7 that the power for the grindstone is provided by an electric drill. Needless to say, I have no connection with the makers of this tool.

The Dormer Drill Information Handbook gives the following general hints on drilling. (I always believe in going to the experts for the best advice.)

1. Keep drills sharp. Frequent resharpening is good economy. It is wasteful to delay resharpening a drill.
2. When a drill is being point ground, ensure that all wear is removed and that the correct point angles are produced.
3. The chuck in which a straight shank drill is held must be of good quality. If the drill slips in the chuck and the feed is automatic, breakage of the drill is inevitable.
4. When driving taper shank drills into sockets, use a soft-face hammer. Make certain that there is a good fit between the taper shank of the drill and the sleeve or socket, otherwise the tang may break.

Fig. 10.7 *Newmartek drill grinding jig.*

5. The work must be held rigid and the machine spindle must have no play.
6. Use recommended lubricants. Take care to ensure that the lubricant reaches the point of the drill.
7. Do not allow the drill flutes to become choked with swarf.
8. Use multi-fluted drills for opening out existing or cored holes. Two-fluted drills are not designed for this purpose.

For anyone just setting up a workshop, and for industry, drills in metric sizes are the best buy. However, fractional drills, that is those in sizes such as ¹⁄₁₆ in., ⅛ in., ¼ in., and ½ in., are very useful and the older engineer still uses them. The third category, number drills, are extremely useful to the model engineer. They range from the smallest, No 80 with a diameter of 0.0135 in., to No 1 with a diameter of 0.228 in. They are particularly useful for tapping and clearance drills for B.A. threads. Letter drills follow on with "A" 0.234 in. up to "Z" with a diameter of 0.413 in.

Drills are made from high speed or carbon steel. The latter are cheaper but not recommended; it is worthwhile paying the little extra for the high speed steel type.

The drilling of most metals is greatly improved by the use of a cutting lubricant, though cast iron should always be machined dry. Brass and bronze can generally be machined dry although some authorities recommend a lubricant. Steel should always be drilled with a lubricant.

By using a lubricant, the drill and the work are cooled and a higher speed can be used; the cutting fluid helps in lubricating the severe rubbing action taking place between the drill lip and the work, and the lubricant helps to wash away the chips and keeps the cutting point clear. Cutting lubricants may consist of pure oil, a mixture of two or more oils, or a mixture of oil and water. Sometimes sulfur is added to give the property of "wetting" the metal with a highly adhesive oil film. The most common type of lubricant is a soluble oil, which, when mixed with water, forms a white, milky solution known as "suds" or "slurry." The supplier's instructions should be followed as regards the proportion of oil and water required. Table 1 gives some guidance on the lubricants to be used when drilling various metals.

Table 1

MATERIAL	CUTTING LUBRICANT
Aluminum and magnesium alloys	Soluble oil or neat cutting oil
Cast iron	Dry
Copper, brass, or bronze	Soluble oil or dry
Mild steel, alloy steel, steel forgings, wrought iron, and monel metal	Soluble oil or sulfurized oil
Stainless steel	Tallow, turpentine, soluble oil, or sulfurized oil

Drilling at the right speed is important. In production engineering, the speed at which drills are run is carefully calculated so that maximum output with optimum drill life is obtained. The fitter working in the general engineering workshop or the model engineer will not go to this length, but better results can be obtained if a speed approximating to the correct one is used. Handbooks published by the drill manufacturers and books on workshop practice give the peripheral speed for drills operating on various materials, and from these, the RPM can be worked out using the following formula:

$$\text{Rev/min} = \frac{\text{Feet/min} \times 12}{3.142 \times D}$$

where feet/min = peripheral speed of the drill in feet per minute and D = the diameter of the drill in inches.

TABLE 2

MATERIAL TO BE DRILLED	SPEED feet/min	SPEED Meters/min
Aluminum and aluminum alloys	100 to 200	30 to 61
Brass	100 to 150	30 to 46
Brass, leaded	100 to 200	30 to 61
Bronze, ordinary	100 to 200	30 to 61
Bronze, high tensile	70 to 100	21 to 30
Gray cast iron	80 to 100	24 to 30
Copper	100 to 150	30 to 46
Monel metal	20 to 50	6 to 15
Free cutting mild steel	80 to 100	24 to 30
High tensile steel	50 to 70	15 to 21

Table 2 gives the recommended speed for various materials.

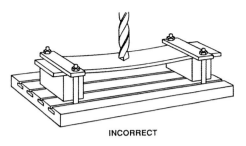

INCORRECT

Fig. 10.8a *Supports too far apart, allowing job to sag.*

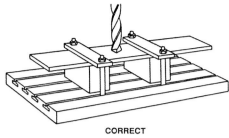

CORRECT

Fig. 10.8b *Job properly secured by clamps.*

The work being drilled must be securely fastened to the drilling machine table. This is particularly important when thin sheet metal is being drilled. When the drill is about to break through the metal, it tends to jam in the hole, causing the metal being drilled to spin around. If the metal is held in the hand, a nasty injury can be caused. Sheet brass is very prone to this trick. A machine vise or clamps should be used to secure the work. Fig. 10.8b shows a piece of metal properly secured by clamps, while Fig. 10.8a shows the supports too far apart, allowing the work to flex and spring.

The small **MACHINE VISE** shown at Fig. 10.9 has jaws 2 ¼ in. wide, and the sliding and the fixed jaws both have V's in a horizontal and vertical position to facilitate the holding of round material. The sliding jaw is constructed so that it can swivel, allowing tapered or irregular shaped work to be held. The side of the jaw against which the screw abuts is semi-circular in shape so that it can be tightened at any angle.

Heavier vises of greater capacity and suitable for use on large drilling machines are obtainable.

V-blocks, described in Chapter 9, are used for holding work on the drilling machine table, especially for cross-drilling shafts. Very often, it is necessary to drill a hole across the exact center of a piece of round material, which is not easy. A small error results in the hole being way off center.

Using a jig is the ideal solution to this problem and a simple one will be described later. If a jig is not available, a line is scribed across the exact center of the end of the shaft in the way described in Chapter 9 on Marking Out. This line is then extended along the shaft. A center-punch mark is then made on this line at the exact distance from the end of the shaft that the hole is to be drilled. The shaft is then clamped in V-blocks, and the scribed line on the end of the shaft is set truly vertical by means of a square. A small center drill is used to enlarge the center punch mark and to provide an accurate start for the drill used to form the hole. See Fig. 10.10.

Fig. 10.9 *Machine vise for use with pillar drill, etc.*

A simple jig for cross-drilling can be made from a short end of ½ in. mild steel square bar as shown in Fig. 10.11. A hole of the same size as the required cross hole is drilled through the square bar, great care being taken to ensure the hole is in the exact center. Another hole is then drilled and reamed lengthwise through the bar to the exact size of the material through which the cross hole is required. Care must be taken to see that this hole is truly central. Having made the jig, the material to be drilled is placed in position in the jig and the cross hole can be quickly and accurately drilled. End location of the cross hole is the only problem, but careful measurement to calculate how much the shaft protrudes from the jig before drilling takes place should not be difficult. If a number of similar holes have to be drilled, some form of stop, to locate the components accurately, should not be difficult to devise. Having made the jig, it should be preserved as it is bound to come in useful on some future occasion. Over the years, a number of these jigs will accumulate and provide a speedy and accurate method of doing a job that can be tedious.

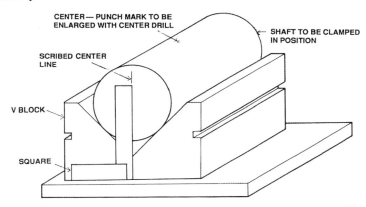

Fig. 10.10 *Method of setting shaft on drilling machine table for drilling cross hole.*

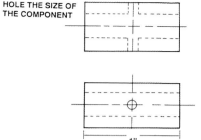

Fig. 10.11 *Jig for drilling cross hole.*

Sometimes it is necessary to drill a hole to meet the side of a previously drilled larger hole. Unless special precautions are taken, there is a danger of the unsupported drill being broken as it breaks through into the large hole. The best way to overcome this problem is to fit a plug, of the same material as that of the work, in the large hole so that the drill is supported throughout its travel (see Fig. 10.12).

A similar difficulty arises when a drill has to be started on an inclined surface, and the safest way, in this case, is to provide a wedged-shape piece of material with such an angle that the drill enters it at a right angle. When the hole has to go right through fairly thin material, it may be possible to use two wedges held together by toolmaker's clamps, as shown in Fig. 10.13. If this is not possible, some other means must be devised to hold the wedge in position.

Where two components have to be held together by nuts and bolts or by studs and nuts, it is a good idea to drill the necessary holes while the two pieces are clamped together with toolmaker's clamps. Where studs and nuts are employed, the tapping drill should be used first. The hole can then be tapped where necessary, while the other hole or holes are opened out with a clearance drill.

Where the exact position of a hole is very important, its location should be marked at the point of intersection of two crossed lines, scribed when the component is being marked out. A light center-punch mark should be made exactly at the point the lines intersect. Accuracy can be checked by using a magnifying glass if there is any doubt that it is not "spot on." If the punch mark is slightly out, it can be corrected by "drawing" the mark over by using the center punch at an angle. From the punch mark, the circumference of the hole must be scribed with the dividers. The circle should then be lightly marked with

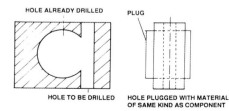

Fig. 10.12 *Method of drilling hole into existing hole.*

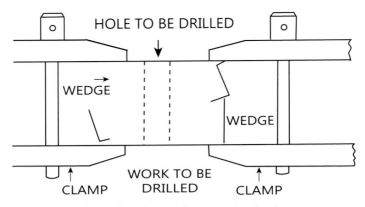

Fig. 10.13 *Drilling a hole on an inclined surface.*

center-punch marks. A drill, smaller than the finished size, should then be used to start the hole. It can then be seen if the drill has wandered and if the full size hole will correspond with the scribed circle. Any small error can be corrected at this stage by cutting a groove with a small diamond point chisel in the direction it is desired the drill should travel. This correction must be made before the drilled hole has reached its full diameter.

If sure that the center-punch mark made initially is accurate, a **CENTER DRILL**, sometimes called a **SLOCOMBE DRILL**, Fig. 10.14, can be used to start the hole and entered up to the full diameter of its shank. The following drill will then be much less likely to wander because its cutting edge will be fully supported.

Fig. 10.14 *Combination center drill, sometimes called a "Slocombe" drill.*

In all cases, where large holes have to be drilled, it is better to start with a small drill and then open out the hole with a larger drill. Drill chucks should be of good quality and drills should be held tightly in them. If the chuck does not hold the drill firmly, then it is likely that the drill will jam in the work. The chuck will go on turning, scoring the shank of the drill and providing conditions where the drill may fracture.

Sometimes it is necessary to countersink a hole to take a countersunk screw. Commercial countersinks are expensive and are inclined to chatter unless run at very slow speeds. Often, the lowest speed available on a drilling machine is not slow enough for this work. The single lip countersink, shown in Fig. 10.15, is easily made from drill rod and avoids the chattering experienced with the expensive commercial tool. The flat cutting edge lies a little above the centerline of the tool. If the tool is made from ½ in. drill rod, it is suggested that the cutting edge be

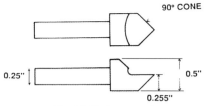

Fig. **10.15** *Countersink.*

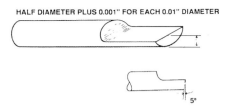

Fig. **10.16** *D-bit.*

made 0.005 in. above the centerline, and this dimension should be checked with a micrometer. The tool should be hardened and tempered by heating to a bright cherry red and then quenched. The tool is then cleaned with fine emery cloth and the shank heated in a small flame until the tip assumes a straw color, when it is again plunged into cold water.

D-BITS of the form shown in Fig. 10.16 will produce very accurate holes and they are easily made. A pilot hole, slightly smaller than the finished hole, is drilled, and then opened out for a short distance with a conventional drill of the finished size. The D-bit is then used to produce an accurate size hole in much the same way as a reamer is used.

The tool is made from drill rod, filed to the shape shown in the drawing, and then hardened and tempered in the same manner as the countersink in the previous paragraph. The flat surface behind the cutting edge must be a little above the centerline of the tool in order to maintain guidance as it is used. The cutting edge is finally honed to a fine finish.

REAMERS are used for finishing drilled holes accurately in size and roundness. Fig. 10.17 shows a hand and machine reamer. The hand reamer,

with which we are at present concerned, is distinguished from the machine one by having a square shank to carry a tap wrench, and a taper lead for guiding it into the hole being reamed.

Obviously, the drilled hole must be slightly, but only slightly, undersize so that there is metal left in the hole for the reamer to clear out. Reamers are intended purely as a finishing tool, so very little metal should be left for the reamer to remove. For example, a $^{31}/_{64}$ in. drill should be used

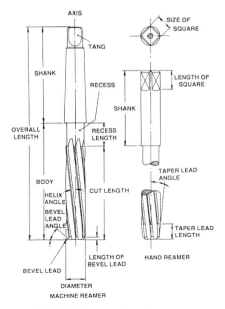

Fig. **10.17** *Hand and machine reamers.*

67

when a ½ in. reamer is going to bring the hole accurately to size.

To use the reamer, the squared shank is gripped in a tap wrench and the tapered end of the reamer inserted in the hole to be reamed. The tool is then turned in a clockwise direction until it has passed fully into the hole. The reamer should be withdrawn from the hole while still turning it in the same direction. The reamer must not be forced into the hole or roughly used, as this may cause damage to the cutting edges. When steel is being reamed, a plentiful supply of lubricant must be used. When a long hole is being reamed, the reamer should be withdrawn from time to time to clear the chips from the flutes.

The machine reamer illustrated differs from the hand type in that it has less taper lead and is fitted with a Morse taper shank so that it can be used in machine tools.

Chapter 11

Screwed Fastenings, Spanners, Screwdrivers, and Pliers

When components have to be dismantled and re-assembled from time to time, various types of screwed fastenings are employed. Where a permanent fixture is required, riveting, soldering, brazing, or welding are used. The permanent fixtures will be reviewed later.

The simplest of the screwed fastenings is the **NUT AND BOLT**, illustrated in Fig. 11.1. The bolt in this case has a hexagon head, and a portion under the head is left plain. The plain section extends beyond the joint face and there must be sufficient thread to allow the nut to seat tightly on the component.

The hexagon **SET SCREW**, Fig. 11.2a, is similar to the bolt but it is threaded right up to the head. Please note that readers in North America call this type of fastener a bolt and, what they call a set screw, those of us on this side of the pond call a grub screw.

The **STUD**, Fig. 11.2b, is threaded at both ends; one end screws into the component while a nut is used at the other end. When a stud is screwed into soft metal, such as aluminum, it is common practice for the thread at that end of the stud to be coarser than the end which carries the nut. This is because the coarser thread is less

Fig. 11.1 *Nut and bolt. Bolt is only threaded part way.*

Fig. 11.2a *Setscrew. Threaded whole length.*

Fig. 11.2b *Stud. Threaded both ends.*

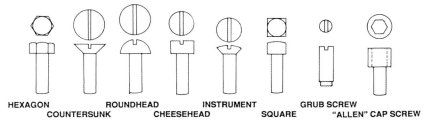

| HEXAGON | ROUNDHEAD | INSTRUMENT | GRUB SCREW |
| COUNTERSUNK | CHEESEHEAD | SQUARE | "ALLEN" CAP SCREW |

Fig. 11.3 *Various types of screw heads.*

likely to strip than if a fine one was used. The fine thread at the nut end provides the clamping pressure needed. Incidentally, a nut with a fine thread is less likely to slacken off than one with a coarse thread.

As well as the hexagon, there are several different types of bolt and screw head to suit various applications. Some of these are illustrated in Fig. 11.3. A component is often counterbored so that a cheese-headed screw is below the surface (see Fig. 11.4). The counter-sunk, or flat head, screw has an angle (90° for metric and 82° for imperial versions) between the sides of its head, and a special countersinking cutter is needed to machine the component into which the screw is going to fit.

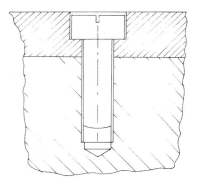

Fig. 11.4 *Counterboring to position cheesehead screw below surface.*

The **GRUB SCREW** (set screw for North American hobbyists) has no head but is turned by a screwdriver fitted in a slot formed in the shank of the screw. It is often used as a locking device, typical examples being the securing of an electric cable in a terminal or the locking of a pulley on a shaft (see Fig. 11.3).

An extremely useful and versatile family of screws are those provided with hexagon sockets, popularly known as "Allen" screws. They are generally made of a high-quality steel and have a hexagon shaped recess into which a special hexagon key fits for tightening or slackening purposes. They are available in cap, grub, and counterbored types. The latter are very useful when fabricating components, as they are flush fitting, very strong, and can be tightened to a much greater degree than the ordinary counterbored screw with a screwdriver slot. A cap screw of this type is shown in Fig. 11.3.

It is often necessary to provide some method of locking a nut or set screw to prevent it slackening off. The degree of security necessary depends on the actual situation. For example, if a big end nut on an internal combustion engine comes adrift, the result is pretty disastrous, while if a nut slackened off on an exhaust manifold on the same engine, it would not be a very serious occurrence.

The most popular positive locking device is a **SPLIT PIN** passing through the slots in a **STANDARD SLOTTED NUT** or a **CASTLE NUT** and through the bolt itself (see Fig. 11.5a and b). Where a group of two or more set screws have to be locked, a wire passing through holes in the heads of all of them, as shown in Fig. 11.5c, may be employed. The locking plate shown in Fig. 11.5d is another common locking device.

Self locking nuts (Fig. 11.5e) have a fiber ring inserted in the center of the nut and a thread is cut in this ring when the nut is tightened. The fiber grips the thread on the bolt so that the nut is unlikely to become slack. They are less effective after they have been removed, and in important locations, they should be renewed once they have been removed.

Two nuts on a stud, as shown in Fig. 11.5f, are often used to prevent slackening off. The first nut is tightened to the correct torque and held by a spanner, while the second one is tightened. When dismantling, two spanners should be used initially, one to hold the bottom nut, while the top one is slackened. If this is not done, there is a tendency for both nuts to move at once, which might cause the thread to strip.

SCREW THREAD NOMENCLATURE

Before discussing the different thread systems, the following definitions of screw thread terms will be useful:

MAJOR OR OUTSIDE DIAMETER is the distance across the thread measured at 90° to the thread axis.

THE CREST is the tip of a thread where the two flanks join.

THE FLANK is the surface of the thread that connects the crest to the root.

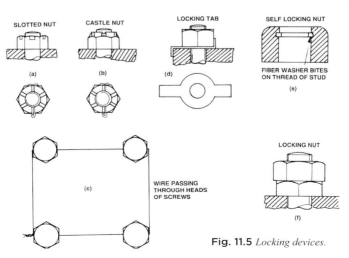

Fig. 11.5 *Locking devices.*

71

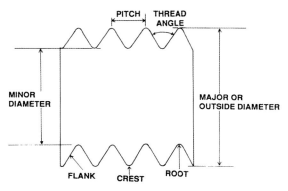

Fig. 11.5g *Thread terms.*

THE ROOT of a thread is the bottom surface joining adjacent flanks.

CORE OR MINOR DIAMETER is the diameter of the thread measured across the roots.

PITCH is the distance from a point of a screw thread to a corresponding point on the next thread, measured parallel to the axis.

LEAD is the distance a screw advances axially in one turn. On a single start thread, used on ordinary bolts and screws, the lead and pitch are identical.

THREAD ANGLE is the angle enclosed by the flanks.

These terms are illustrated in Fig. 11.5g.

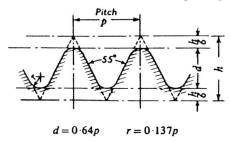

$$d = 0 \cdot 64p \qquad r = 0 \cdot 137p$$

Fig. 11.6 *Whitworth and British Standard Fine threads.*

SCREW THREAD SYSTEMS

Imagine the chaos that would arise if every manufacturer had its own thread system; repairing machinery would be a nightmare. Sir Joseph Whitworth, in 1841, realized that the work of engineers was greatly impeded by the lack of a thread standard, and he instituted the system which bears his name. The form of this thread, known as the **BRITISH STANDARD WHITWORTH, (BSW)** is shown in Fig. 11.6. For some purposes this thread is too coarse and the **BRITISH STANDARD FINE** was introduced. This thread has the same form as the BSW, but with a finer pitch.

In 1945, following a series of conferences between Britain, America, and Canada, the **UNIFIED THREAD** was introduced. There are two types, the **UNIFIED COARSE (UNC)** and the **UNIFIED FINE (UNF)**. The thread takes the form shown in Fig. 11.7, and it will be seen that the angle of this thread is 60° as opposed to the 55° of Whitworth.

Spanners for the Whitworth sizes are designated according to the bolt diameter; for example, the jaws of a ¼ in. Whitworth spanner fit a nut on a ¼ in. bolt but the jaws of the spanner are a little under ½ in. apart. Spanners for the unified bolts are named according to the distance across the flats of the nuts which they fit, and because of this, they are known as "AF" ("across flats") spanners.

In 1965, the President of the Board of Trade announced that British Industry should progressively adopt metric units, and following this announcement, the ISO screw was introduced. It has the same form as the Unified thread, illustrated in Fig. 11.7 but, of course, the diameters are in metric sizes.

The **BRITISH ASSOCIATION (BA)** thread has been in use in the UK for many years for small screws, especially for electrical and instrument work. It is very popular with model engineers and is particularly suitable for their hobby. It has the form shown in Fig. 11.8, and the different diameters are designated as numbers ranging from "O," the biggest in the range with a diameter of 6mm, to the smallest, "23" with a diameter of 0.33mm. The usual set of taps and dies covers the sizes from 0 to 10. The number 10 screw has a diameter of 1.7mm (0.0669 in.), plenty small enough for most purposes.

The **BRITISH STANDARD PIPE (BSP)** thread is used chiefly for gas and water pipes, but is sometimes used for other purposes where a fine thread is needed on large diameter work.

The **BRITISH STANDARD BRASS** thread is used for brass tubing, gas fittings, and general brass work. The thread is of Whitworth form and all sizes have 26 threads per inch.

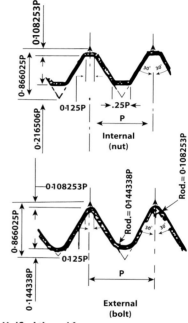

Unified thread from
- Basis of all modern vee threads
- UNC – unified coarse
- UNF – unified fine
- UNS – unified special
- ISO – metric

Fig. 11.7 *Unified threads.*

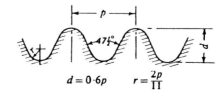

$$d = 0 \cdot 6p \qquad r = \frac{2p}{\Pi}$$

Fig. 11.8 *British Association (BA) threads.*

Fig. 11.9 *Torque wrench.*

There is also a **MODEL ENGINEER** thread, which has the same form as the Whitworth thread. Diameters up to ¼ in. have 40 threads per inch (TPI). Above that size, 32 TPI is the norm, although 40 TPI are found sometimes on bolts above that size.

It is important that nuts and bolts are tightened to the correct torque. Torque, as far as nuts and bolts are concerned, means the pull on the spanner that turns, or tries to turn, the nut or bolt. When a spanner is used to tighten a nut, it acts as a lever and the torque equals the force applied to the spanner multiplied by the length of the spanner, the answer being given in pounds force inches (lbf ins), pounds force feet (lbf ft), kilograms force meters (Kg fm) or Newton meters (Nm), depending on which system is used.

Many workshop manuals give the correct torque for tightening particular bolts, and in these cases a torque wrench is used. A torque wrench has an adjustment that can be set, so that as a bolt is tightened and the required torque is reached, a loud metallic "click" is heard. On hearing this "click," one is assured that the bolt has been correctly tightened and no further action is needed. A torque wrench is shown in Fig. 11.9.

Where the bolt or stud is too small for a torque wrench to be used, or one is not available, the correct tightness has to be judged. The pull on the spanner should be smooth and on no account should it be jerked. It is important that the correct length spanner is used. Remembering that torque is the force applied multiplied by the length of the spanner, it will be seen that lengthening the spanner increases the torque even though the force applied remains the same.

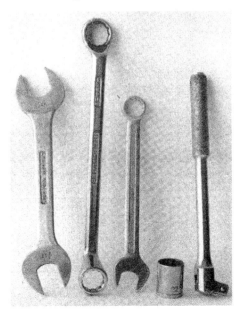

Fig. 11.10 *Different types of spanners.*

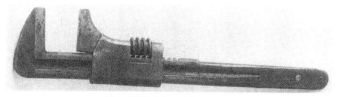

Fig. 11.11 *Adjustable spanner.*

SPANNERS

There are various types of spanners, or wrenches, to fit all the bolts and nuts described in this chapter. Fig. 11.10 shows those in general use. On the extreme left is the **OPEN-ENDED** type. All spanners should be a good fit on the nuts on which they are used and this is particularly important with the open-ended type. Because they only engage on two of the flats on the nut, they are prone to slip off the nut when in use and, of course, if they are not a good fit, this is much more likely to occur. Barked knuckles and rounded corners on nuts are the inevitable results of this malpractice.

A **BOX END SPANNER** is shown in Fig. 11.10, next to the open-ended one. On these spanners, the holes which accommodate the nuts are bi-hexagonal in shape, i.e., they have twelve corners and are, therefore, easier to use in a confined space, as they need only be turned one-twelfth of a turn before they can be removed from the nut and reengaged for a fresh pull. They also have the advantage of fully surrounding the nut that renders them less likely to slip off.

The **COMBINATION SPANNER**, to the right of the box end spanner in Fig. 11.10, has a ring at one end and is open ended at the other end. Open-ended and box-end wrenches have different sizes at each end but combination spanners are supplied with both ends the same size.

Next, in Fig. 11.10 is the **SOCKET SPANNER**, a short rigid tube having on one end an internal bi-hexagonal recess. The other end has a ½ in. square hole to take a handle. (Spanners with larger or smaller square holes are available for larger and smaller spanner sizes) A variety of handles is available, including a brace or speed wrench, a hinged handle, and a ratchet. Extension bars and a universal joint are also generally included in a set of this kind. A hinge-type handle is shown adjacent to the socket in Fig. 11.10.

An **ADJUSTABLE SPANNER** is shown in Fig. 11.11. These have one fixed and one sliding jaw. Their use is not recommended if a fixed spanner of the correct size is available.

ALLEN KEYS are small "L"-shaped hexagon bars that fit into the recess of hexagon socket screws that were described earlier in this chapter. See Fig. 11.12.

Fig. 11.12 *Allen key.*

Fig. 11.13 *Box spanners.*

BOX (or **TUBULAR**) **SPANNERS** (Fig. 11.13) have been largely superseded by the socket wrench, but they are still useful for special jobs where a nut has to be tightened in a deep recess, or where the stud, for some special reason, protrudes a long way out of its nut.

C-SPANNERS, which go by the name of "spanner wrenches" in NA, are seen in Fig. 11.13a and are used on the fluted nuts sometimes fitted to pump or piston rod glands.

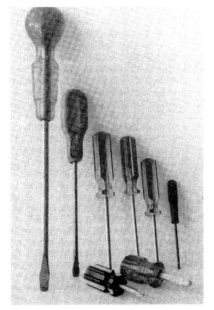

Fig. 11.14 *Various types of screwdrivers.*

SCREWDRIVERS

Good-quality screwdrivers are made from chrome alloy steel, suitably hardened and tempered. They vary in length from the "stubby," made especially for working in confined spaces, to the very large ones, about 24 in. (610mm) long.

Fig. 11.13a *C-spanner.*

A variety of handles is available, but probably the best types for the engineer are those with tough, insulated plastic handles. Fig. 11.14 shows a selection of various types.

Special screwdrivers are made for the Pozidriv and Phillips type screw heads. Although these are similar, in that they fit screws with a cross type head instead of a single slot, they are not identical. The Phillips screwdriver will only give a poor fit on a Pozidriv screw and vice versa.

A screwdriver, like a spanner (for actually it is an internal spanner), should do its work by being applied, and then turned in the required direction with the screw as its axis. If it is not properly formed and of the correct shape, it will require a great amount of pressure while being turned. Many amateurs experience troubles because their screwdriver is incorrectly sharpened and/or they are using the wrong type for a particular job.

Fig. 11.15 illustrates correct and incorrect use of screwdrivers. At (a), the

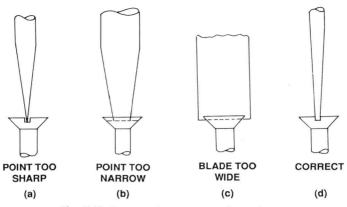

POINT TOO SHARP (a) POINT TOO NARROW (b) BLADE TOO WIDE (c) CORRECT (d)

Fig. 11.15 *Correct and incorrect use of screwdrivers.*

screwdriver has a sharp point and will be inclined to slip out of the screw slot. The screwdriver at (b) is too narrow for the screw and full pressure will be impossible to apply. At (c), the screw-driver is too wide, and when undoing a countersunk screw, the edges will foul the sides of the work. At (d), the correct screwdriver, with the right shaped blade, is shown.

It is a strange fact that a long screwdriver moves a screw more easily than a short one, even though the mechanical advantage

Fig. 11.16 *Ratchet screwdriver.*

regarding the size of the handle and width of the point is the same. This may be because the operator can more easily apply his strength with the big tool than with a small one.

Ratchet screwdrivers, sometimes called a Yankee screwdriver, speed up the work,

especially on repetition work One working on the Archimedean principle, which will insert screws using a downward movement of the handle, is illustrated in Fig. 11.16.

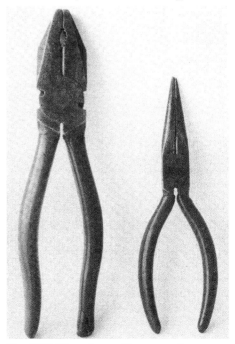

Fig. 11.17 *Side-cutters and long-nose pliers.*

77

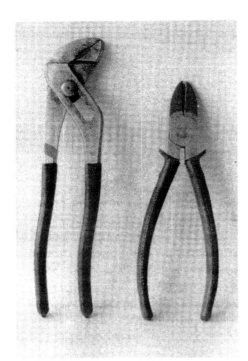

Fig. 11.18 (above) *Water pump pliers and side-cutters.*

PLIERS

Strictly speaking, pliers should not appear in a chapter relating to screwed fastenings, as they should never be used to turn bolts or screws. It does seem convenient, however, to deal with them here, as they are classed as small tools which "live" in the toolbox with wrenches and screwdrivers.

A large variety of pliers is available but three or four pairs are all that are really necessary for ordinary work. The side-cutting and long-nose shown in Fig. 11.17 are the most useful types. The side-cutting pliers have serrated jaws and, at the inner end of the jaws, cutting edges. A shear type of wire cutter is arranged at each side of the rivet on which the pliers swivel.

The **WATER PUMP** pliers, with a slip joint, is shown in Fig. 11.18 together with a pair of side-cutters, which are a useful addition to the tool kit. **STILLSONS** and "**FOOTPRINTS**," shown at Fig. 11.19, are used for pipework and turning other round materials. They should never be used on nuts or bolts as they have serrations that would cause serious damage.

Fig. 11.19 *Stillsons and "Footprints."*

Chapter 12

Taps and Dies

One of the basic skills an engineer has to acquire is producing external and internal threads by the use of taps and dies. Careless or improper use of these tools can cause endless trouble, as will be explained later. **TAPS** for producing internal threads are supplied in sets of three: **TAPER, SECOND, AND PLUG**. (Fig. 12.1). The taper tap has fully defined screw threads only at the top end of the shank. The formation of the thread diminishes toward the lower end. The second tap is similar to the taper but it has a longer length of fully defined thread with only a comparatively short length of taper. The plug tap has fully defined threads for practically its full length. The top end of the shank of the tap is square to accommodate a **TAP WRENCH**, examples of which are shown in Fig. 12.2.

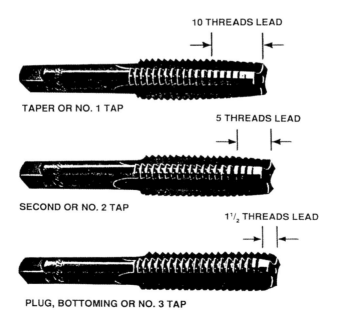

10 THREADS LEAD

TAPER OR NO. 1 TAP

5 THREADS LEAD

SECOND OR NO. 2 TAP

1$\frac{1}{2}$ THREADS LEAD

PLUG, BOTTOMING OR NO. 3 TAP

Fig. 12.1 *Taper, second, and plug taps.*

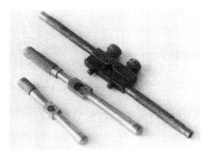

Fig. 12.2 *Tap wrenches.*

For tapping, a hole must first be drilled to a diameter slightly greater than the core diameter of the thread. Published tables give recommended tapping drill sizes for various threads, and the tables shown are taken from Neill Tool Users' Handbook, reproduced here by kind permission of Neill Tools Ltd. The tapping sizes recommended are for use only on steel, and will give approximately 75% depth of thread. Deviations from stated tapping drill sizes will be necessary depending on the nature of the material and the requirement of the finished work.

It is unlikely that drills of the exact metric size given in the tables will, in all cases, be available. Often it will be found that a fractional or number drill, very near the recommended metric size, will do very well. For example, a 2.7mm drill is recommended for tapping 5 B.A. A number 37 drill is very near this size and will do the job very well. The torque required to turn the tap will depend on the size of the tapping drill: the smaller the drill, the greater the effort required. This is an important consideration, especially with the smaller taps, if breakages are

to be avoided. It may be better to use a drill at least one size larger than the one recommended if the resistance encountered when turning the wrench is excessive.

In the smaller sizes, likely to be used by the model engineer, taps often break off in the work and this is very frustrating, for not only will a new tap need to be bought, but the broken end of the tap has to be removed from the workpiece. This can be very difficult and, sometimes, almost impossible. The tap is hard, so it cannot be drilled out and some other means has to be devised. If part of the broken tap protrudes from the work, it may be possible to grip it with a pair of pliers and screw it out, but I have never had much luck with this procedure. Some textbooks say it may be possible to screw the broken end out with light blows with a hammer and punch, but here, again, I have never found this to work. Sometimes, if the right equipment is available, the tap can be heated and softened and then drilled out. This is another chancy technique and it is far better to be extra careful and avoid breaking a tap than to be faced with the task of removing a broken one.

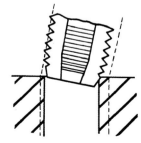

Fig. 12.3 *An oversize and bell mouth tapped hole will result if tap does not enter the hole squarely.*

Tapping and Clearance Drill Chart

These tapping drills are recommended for use on steel only and will give approximately 75% depth of thread.

Deviations from stated tapping drill sizes will be necessary, depending on the nature of material and requirements of finished work.

BA

Dia	Tapping	Clearance	Dia	Tapping	Clearance
0	5.10mm	6.20mm	7	2.10mm	2.60mm
1	4.50mm	5.50mm	8	1.80mm	2.30mm
2	5/32"	4.90mm	9	1.50mm	2.00mm
3	3.50mm	4.20mm	10	1.40mm	1.80mm
4	3.00mm	3.70mm	11	3/64"	1.60mm
5	2.70mm	3.30mm	12	1.05mm	1.40mm
6	3/32"	2.90mm			

BS Fine

Dia	Tapping	Clearance	Dia	Tapping	Clearance
3/16"	4.00mm	4.90mm	1/2"	7/16"	33/64"
7/32"	4.70mm	5.70mm	9/16"	1/2"	37/64"
1/4"	5.40mm	6.50mm	5/8"	9/16"	41/64"
9/32"	6.20mm	7.40mm	11/16"	5/8"	45/64"
5/16"	6.90mm	8.20mm	3/4"	43/64"	49/64"
3/4"	21/24"	9.80mm	13/16"	47/64"5	3/64"
7/16"	9.80mm	29/64"	7/8"	51/64"	57/64"
			1"	29/32"	1 1/64"

BS Whitworth

Dia	Tapping	Clearance	Dia	Tapping	Clearance
1/16"	3/64"	1.70mm	3/8"	8.00mm	9.80mm
3/32"	1.90mm	2.50mm	7/16"	9.30mm	29/64"
1/8"	2.60mm	3.30mm	1/2"	27/64"	33/64"
5/32"	3.20mm	4.10mm	9/16"	31/64"	37/64"
3/16"	3.80mm	4.90mm	5/8"	17/32"	41/64"
7/64"	4.60mm	5.70mm	3/4"	31/32"	49/64"
1/4"	13/64"	6.50mm	7/8"	49/64"	57/64"
5/16"	6.60mm	8.20mm	1"	7/8"	1 1/64"

(continued on next two pages)

As the action of tapping raises a swelling around the hole, it is generally advisable to countersink a hole slightly before starting to tap it.

Having drilled and countersunk the tapping hole, the taper tap is fixed to the tap wrench and started in the hole. In order to avoid an oversize and bell-mouth tapped hole, caused by misalignment of the tap (Fig. 12.3), it is necessary to check that the tap is entering the hole squarely. This may be done with a small square as is shown in Fig. 12.4.

Tapping and Clearance Drill Chart *(continued)*
Unified

Diameter	Tapping		Clearance
	UNC Coarse	**UNF Fine**	
0	–	$\frac{3}{64}$"	$\frac{1}{16}$"
1	1.45mm	1.5mm	1.95mm
2	1.7mm	1.8mm	2.25mm
3	$\frac{5}{64}$"	2.05mm	2.6mm
4	2.2mm	2.3mm	2.95mm
5	2.55mm	2.6mm	3.3mm
6	2.7mm	2.9mm	$\frac{9}{64}$"
8	3.4mm	3.5mm	4.3mm
10	3.8mm	4mm	4.9mm
12	4.4mm	4.6mm	5.6mm
$\frac{1}{4}$"	$\frac{13}{64}$"	5.4mm	6.5mm
$\frac{5}{16}$"	6.5mm	6.9mm	8.2mm
$\frac{3}{8}$"	$\frac{5}{16}$"	8.5mm	9.8mm
$\frac{7}{16}$"	9.3mm	$\frac{25}{64}$"	$\frac{29}{64}$"
$\frac{1}{2}$"	$\frac{27}{64}$"	$\frac{29}{64}$"	$\frac{33}{64}$"
$\frac{9}{16}$"	$\frac{31}{64}$"	13mm	$\frac{37}{64}$"
$\frac{5}{8}$"	$\frac{17}{32}$"	14.5mm	$\frac{41}{64}$"
$\frac{3}{4}$"	16.5mm	$\frac{11}{16}$"	$\frac{49}{64}$"
$\frac{7}{8}$"	$\frac{49}{64}$"	20.5mm	$\frac{57}{64}$"
1"	$\frac{7}{8}$"	23.5mm	$1-\frac{1}{64}$"

The bench drilling machine may be used for tapping holes if it is provided with a handle to turn the spindle while the tap is held in the chuck. This ensures that the tap is kept square with the work. Some years ago, I built a drilling machine, but when I acquired a larger and more sophisticated machine, I adapted the small one for tapping holes (see Fig. 12.5).

Tap and dies should be lubricated when used on all materials except cast iron. A commercial grease is made especially for this purpose and gives excellent results. I have never felt the need to use a lubricant when working on brass, but one leading manufacturer of screwing tackle makes the following recommendations:

METAL	LUBRICANT
Steel	Sulfurized oil
Cast iron	Dry or light soluble oil
Aluminum	Paraffin or mineral oil
Brass and copper	Paraffin and lard oil
Phosphor bronze	Lard or light oil

ISO metric coarse pitch threads

Dia	Tapping	Clearance	Dia	Tapping	Clearance
M1	0.78mm	1.1mm	M11	9.6mm	11.5mm
M1.1	0.88mm	1.2mm	M12	10.4mm	13mm
M1.2	0.98mm	1.3mm	M14	12.2mm	15mm
M1.4	1.15mm	1.5mm	M16	14.25mm	17mm
M1.6	1.3mm	1.7mm	M18	15.75mm	19mm
M1.8	1.5mm	1.9mm	M20	17.75mm	21mm
M2	1.65mm	2.2mm	M22	19.75mm	23mm
M2.2	1.8mm	2.4mm	M24	21.25mm	25mm
M2.5	2.1mm	2.7mm	M27	24.25mm	28mm
M3	2.55mm	3.2mm	M30	27mm	31mm
M3.5	2.95mm	3.7mm	M33	30mm	34mm
M4	3.4mm	4.3mm	M36	32.5mm	37mm
M4.5	3.8mm	4.8mm	M39	35.5mm	40mm
M5	4.3mm	5.3mm	M42	38mm	43mm
M6	5.1mm	6.4mm	M45	41mm	46mm
M7	6.1mm	7.4mm	M56	51mm	58mm
M8	6.9mm	8.4mm	M60	55mm	62mm
M9	7.9mm	9.4mm	M64	58mm	64mm
M10	8.6mm	10.5mm	M68	62mm	70mm

There are also ISO METRIC FINE PITCH THREADS but as these are many and varied, rather than present in table form it is suggested that the tapping drill is derived as follows:

Tapping Drill = Outside dia.—Pitch e.g., 4.0mm dia 0.5mm pitch

Tapping Drill = 4.0 – 0.5 = 3.5mm

Clearance Drills as per table for COARSE PITCH THREADS.

Fig. 12.4 *Checking that tap is entering work squarely by using a small square.*

When the taper tap is felt to have started its work and its squareness has been checked, cutting of the thread can proceed. The tap should not be turned continuously, but at about every half turn, it should be reversed slightly to clear the thread. If any undue stiffness is encountered, no force whatsoever should be used but the tap must be very carefully eased backwards to clear it.

When a blind hole is being tapped, the tap should be removed occasionally to clear the metal cuttings from the bottom of the hole. If the hole is a straight through one, a reduction in resistance will indicate that the taper tap is cutting a full thread. It can be removed from the hole, which may be finished off with a second tap.

Fig. 12.5 *Drilling machine adapted for tapping holes.*

With blind holes, resistance will be felt when the taper tap reaches the bottom of the hole. The tap is then removed and the second one taken down as far as possible. Finally, the plug tap is used to cut the thread at the bottom of the hole.

STOCKS AND DIES. The tool used for cutting external threads is known as a **DIE** and the device for holding the die is called a **STOCK**. The most common type of die is the split circular die shown in Fig. 12.6, fitted in a stock. The split permits limited adjustment in the size of the thread cut by the die. The ring portion of the stock into which the die fits contains three screws. The center screw has a tapered end that fits in the split in the die. The other two screws press on to the periphery of the die. By slackening the two outer screws and tightening the center one, the die is expanded so that a shallower cut is taken. If the center screw is slackened and the outer ones tightened, then the die is squeezed in and a deeper thread is cut. Although the amount of adjustment is slight, it is extremely effective in ensuring a good fitting thread.

Cutting an external thread with a die is a similar operation to using a tap and tap wrench but there is the problem of starting the die absolutely square. Generally, only the first two threads of the die are chamfered and this does not give as much assistance as the long taper on a taper tap. Some stocks and dies are fitted with detachable collets, which ensure the thread is cut squarely. There is a selection of collets and it is only necessary to find

Fig. 12.6 *Circular stocks and dies.*

one of the correct size for the job in hand and place it in the stock, and it will keep the die square with the work.

The following method of using dies is suggested:

1. Check that the rod diameter is correct and chamfer the end to give the die a start.
2. Spread the die with the center screw so that a shallow cut is taken at first.
3. Check the die is square with the rod to prevent a "drunken" thread being cut.
4. Close the die to the finish size by slackening the center screw in the die stock and tightening the other two. Re-cut the thread to the correct depth.
5. Reverse the die stock every half turn to clear the swarf.
6. Dies must be lubricated in the same way as taps.

Sometimes it is necessary to cut a thread close up to the head of a bolt. Because the leading edge of a die is tapered a full thread is not cut up to the head of a bolt. This can be remedied, to a limited extent, by reversing the die in the stock after the thread has been cut, and then re-applying it to the bolt. The other and probably better way is to have a recess machined close to the head of the bolt, as is shown in Fig. 12.7. This recess should be at the same depth as the thread. If there is no objection to using a washer on the bolt, it does not then matter that the thread is not cut to its full depth for its complete length.

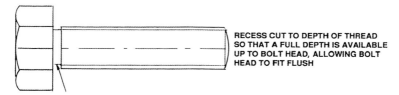

RECESS CUT TO DEPTH OF THREAD
SO THAT A FULL DEPTH IS AVAILABLE
UP TO BOLT HEAD, ALLOWING BOLT
HEAD TO FIT FLUSH

Fig. 12.7

Chapter 13

Riveting

Where a permanent connection in metal is required, riveting is a satisfactory method of making a joint and is extensively used in boiler work, shipbuilding, on girders for building work, and in civil and general engineering.

On models, as well as making a secure joint, the rivet heads must look right if a model is to appear authentic.

Rivets are made from any of the metals that are malleable and generally are of a similar material to that being joined. They are commonly made of steel, copper, brass, and aluminum. Rivets made from the three latter metals are easily clenched because the metals are relatively soft; steel rivets may be riveted up hot or cold.

Fig. 13.1 shows the standard types of rivet heads. The round head is the most commonly used; it gives maximum strength and is easily formed. Tinsmiths use the flat head rivet, and I have found these, with a shank of about ⅛ in. diameter and ¼ in. long, to be very useful, not so much for model making but for general work in the workshop, especially when welding facilities are not available.

Where the round head rivet cannot be used because the projecting head is an inconvenience, the countersunk type enables a flush finish to be attained. The double countersunk, shown in Fig. 13.2, can be made from wire or rod, each end being riveted over and then filed flush.

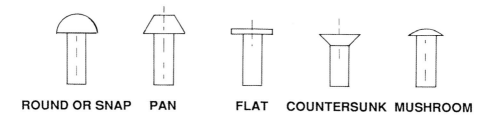

ROUND OR SNAP   **PAN**   **FLAT**   **COUNTERSUNK** **MUSHROOM**

Fig. 13.1 *Standard rivet heads.*

It is not necessary to have the same type head at each end of the rivet. For example, a round head rivet can be clenched with a countersunk finish if a smooth surface is required on one side of the plates being joined.

Riveted joints take various forms. Fig. 13.3a shows a lap joint and also illustrates the ideal spacing of rivets and the amount of overlap of the joint. Fig. 13.3b shows a butt joint and Fig. 13.3c a double riveted butt joint with cover plates on each side.

When joining two pieces of plate or sheet material, the position of the rivets should be accurately marked out, as irregularly spaced rivets spoil the appearance of the work. Fig. 13.4 shows the rivets on the firebox end of a portable steam engine in the course of restoration, and it can be seen that, as well as making a sound joint, the rivets look right.

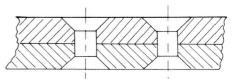

Fig. 13.2 *Double countersunk rivets can be used where a flush finish is necessary.*

A line indicating the position from the edge of the work that the rivets will occupy can be scribed with odd leg calipers (Jennies) and the pitch centers of the rivets marked off with dividers and then with a center punch.

If possible, the plates forming the joint should be clamped together with the top plate marked out for the holes as described. The holes may be drilled through both plates at once so that the holes are bound to align, making the insertion of the rivets easy. The order in which the holes are drilled will depend on the nature of the

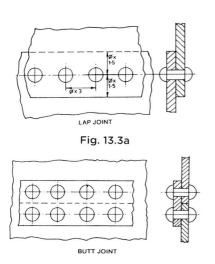

LAP JOINT

Fig. 13.3a

BUTT JOINT

Fig. 13.3b

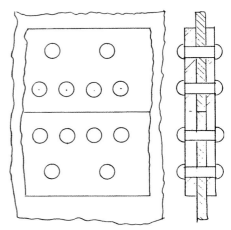

Fig. 13.3c *Double riveted butt joint with cover plates both sides.*

Fig. 13.4 *Rivets on the firebox of a portable steam engine.*

a round head rivet and the diameter of the rivet, plus the thickness of the materials for the countersunk one.

Rivet holes must be of the correct size. If the holes are too large, the rivets will tend to bend over instead of expanding to form a nice tight fit. If the holes are too small, the rivets can only be inserted with difficulty and the edge of the hole may shear the sides of the rivets so that the head will be unable to bed properly (see Fig. 13.5).

Rivets may be clenched in two ways, with a round head or countersunk. When a round head is required, the procedure illustrated in Fig. 13.6 is followed. A tool with a rounded recess in its head, to fit the head of the rivet, is gripped in the vise. The closing tool, basically a punch with a hole in its end slightly larger than the shank of the rivet, is used to ensure the rivet head is tight against the work and that the two pieces of plate have no gap between them. The shank of the rivet is then given a few well-directed blows with the hammer, the aim being to swell the rivet shank in the hole and, at the same time, rivet over the tail. Finally, the rivet is finished to shape with the rivet snap. During all these operations, the rivet head remains seated on the tool gripped in the vise.

Many riveting failures arise through a beginner over-hammering a rivet, that is employing a number of slight, feeble blows when two or three heavier ones would suffice. The rule is that the fewer the number of blows delivered to close down a rivet, the more closely will it secure the

work. In some cases it may be advisable to drill two holes at the ends of the work and secure the plates together with nuts and bolts. Further holes can then be drilled and rivets inserted and clenched. When the plates are fully secure, the bolts can be removed and rivets inserted in their place.

It is important that the correct size rivet is chosen. A general guide to rivet diameter is to make $d = 1\frac{1}{2} t$, where $d =$ rivet diameter and $t =$ the thickness of the plates to be joined. For thin plate, this rule may give too small a rivet so the rule must be used with discretion.

The correct length rivet must be selected; if it is too long or too short, the rivet snap cannot form the correct shaped head. As a general rule, the length of a rivet should be $1\frac{1}{2}$ times the diameter, plus the thickness of the materials being riveted for

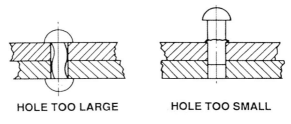

HOLE TOO LARGE   **HOLE TOO SMALL**

Fig. 13.5 *Correct size holes are important.*

joint and the softer and more ductile the metal remains.

In most of the work done by model engineers and also in general engineering, small rivets are riveted while cold. It is only in the larger sizes that rivets are clenched hot. The advantages of hot riveting lie in the fact that less violence is needed to form the head, so that the plating is not so severely stressed, and that the contraction of the hot rivet, as it cools down, tends to more effectively close up the joint.

If a flush finish is required, each rivet hole may be slightly countersunk and the rivet shank expanded into the countersunk portion of the work. Too deep a countersink will entail unnecessary hammering to expand the rivet and should be avoided. After the rivet has been firmly clenched, it can be filed flush and it should then be almost invisible. Care must be taken not to bruise the metal around the hole when riveting, as this can be very unsightly.

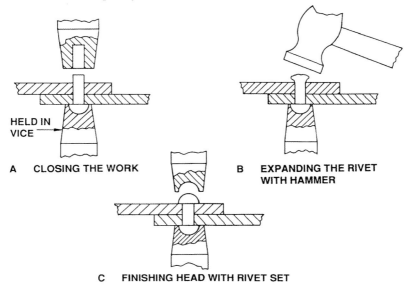

A   CLOSING THE WORK   **B   EXPANDING THE RIVET WITH HAMMER**

C   FINISHING HEAD WITH RIVET SET

Fig. 13.6 *Stages in riveting.*

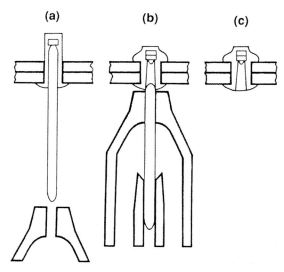

Fig. 13.7 *Pop rivets: (a) rivet inserted; (b) closing tool applied and rivet clenched; (c) further pressure breaks wire.*

The holding-up tool, colloquially known as a "dolly," the rivet snap, and the closing tool can all be easily made from drill rod. The closing tool only needs a hole drilled in its end deep enough to take the rivet shank, but for the other two, it is necessary to make a special form tool to machine the recess to take or form the rivet head. Alternatively, if a hardened steel ball of the same diameter as the rivet head is available, this can be used to forge the right shape. A hole should be drilled in the end of a piece of drill rod of a suitable diameter for a depth slightly less than half the diameter of the ball. The steel should then be heated, the ball placed in the hole, and given a smart blow with the hammer. This treatment should provide a recess the same shape as the rivet head. All three tools must be hardened and tempered.

The support afforded to the rivet during riveting is most important because if it is not of a solid, unyielding nature a satisfactory job cannot be achieved.

"Pop" rivets, illustrated in Fig. 13.7, where the design of the component prevents a dolly being used to support the rivet, are used extensively in industry. The rivet is specially constructed with a wire running through its center. This wire is pulled by a special closing tool to clench the rivet.

Bifurcated rivets, where the shank is divided into two legs, and hollow rivets are used for light work but these cannot be expected to give as satisfactory a joint as a solid one.

Chapter 14

Soft Soldering

Editor's Note: While many aspects of metalworking have changed very little since this book was first published in 1988, the equipment and materials used for soldering and welding have advanced a great deal. The basics remain the same, but the reader is advised to consult with a reputable supplier as to the proper materials required for the task at hand.

Soft soldering is a very useful method of making joints in most metals and particularly in copper, tin, brass, steel, and tinplate and for joining electric cables. Aluminum cannot be soldered using the conventional methods. Soldering is not a satisfactory medium where much strength is required or where the joint is subject to vibration or heat, as a soldered joint is comparatively weak and solder has a low melting point. Where a permanent joint is required and the joint must stand up to heavy loads or high temperature, riveting, brazing, or welding should be used.

The solder I am discussing goes under the general name of **TINMAN'S SOLDER**, an alloy of tin, lead, and antimony. It is sold in sticks about 400mm (16 in.) long and roughly triangular in section. British Standard 219:1959 gives a list of solders, their content, and other properties, and the following table sets out those in general use.

Grade "C" is suitable for general work, but Grade "K," which has minimal antimony content, is recommended for fine work, particularly model engineering projects. It has a shorter melting range,

BS219	Tin%	Lead%	Antimony%	Melting Range Solid/Liquid °C		Properties and Use
A	65	34.4	0.6	183	185	Electrician's solder for wiring electrical components liable to damage if heated.
B	50	47.5	2.5	183	204	Coarse Tinman's solder, a cheap grade for general use.
C	40	57.8	2.2	183	227	General work, cheaper than 'B'.
F	50	49.5	0.5	183	212	For electrical work and zinc.
K	60	39.5	0.5	183	188	A free running, quick setting solder for high class work.

penetrates joints more readily, and is easy to use. It is slightly more expensive than the other grades but is worth the extra outlay.

When a good soldered joint is made, a section through the joint appears as in Fig. 14.1. The solder reacts with the parent metal to form an amalgam. For this intimate union to take place between the metal and the solder, the surfaces of the joint must be scrupulously clean. If molten solder is placed on a dirty surface, it will not wet the surface of the metal but will remain in a globular state. The dirt and an oxide film prevent the close contact necessary for the solder to spread over and wet the surface.

Cleaning may be done with emery or sandpaper, by filing, or with wire wool. One authority warns that emery cloth should not be used on soft metals, such as copper, unless the surface is subsequently scraped all over, because particles of emery become embedded in the surface and may prevent the solder adhering properly. I have never experienced this trouble but pass the information on for what it is worth.

When heated, the surface of the joint will be attacked by the oxygen in the air, and a flux is used to keep the air away to prevent oxidation and to destroy the oxides by chemical action. An illustration, taken from a very old booklet published by the Tin Research Institute, reproduced in Fig. 14.2, shows how the flux is displaced by the molten solder.

There are two kinds of fluxes, those that not only protect the surface but play an active part in cleaning it and those that only protect a previously cleaned surface. The first type is the more efficient, zinc chloride being the one generally used. This is made by dropping zinc cuttings into spirits of salt (hydrochloric acid) contained in a stone jar. This zinc bubbles up and dissolves and the solution is then ready for use. When I was an apprentice, I was sent to the local chemist for the spirits of salt. In those days, there were no domestic refrigerators but most homes had a "meat safe," a cupboard with a perforated zinc door and sides to allow for ventilation but to keep out the flies. Scrap ends of the perforated zinc were always available

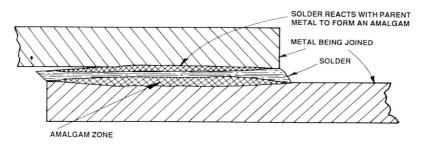

Fig. 14.1 *Section through a soldered joint shown in exaggerated form.*

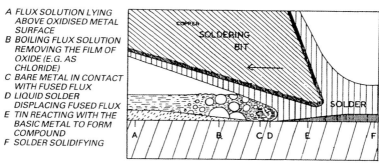

Fig. 14.2 *Diagrammatic representation of the displacement of flux by molten solder.*

and small pieces were dropped into the spirits of salt in a lead container kept for this purpose.

I only described this method as a matter of interest as there is now a commercial product on the market, Baker's Fluid, which can be bought relatively cheaply and which is all ready for use. I keep a small quantity in a shallow china jar, which originally contained meat paste, and which is ideal for this purpose.

There is a big snag with this acid-type flux: it is highly corrosive and joints made with its use must be thoroughly washed with hot water when completed, or dipped in a weak alkaline solution, such as ammonia, to neutralize the acid. There is no guarantee that all the flux will be expelled when the solder runs into the joint and any remaining therein could cause serious corrosion. The vapor given off when using the acid-type flux is objectionable and does cause rusting on any ferrous metals nearby. If at all possible, and weather permitting, I do my soldering outside the workshop in the open air, well away from my machine tools. In

spite of these drawbacks to the acid-type flux, it is the most effective in use, and if its corrosive nature is always borne in mind, and suitable precautions taken, it is very suitable for many jobs.

The second type of flux has only very moderate cleansing properties but does prevent oxides forming and is effectively non-corrosive. It must be used where washing off is not practical or where subsequent corrosion cannot be tolerated. This type of flux should always be used, for example, for joining electric cables. Resin, turpentine, and Vaseline (petroleum jelly) form the basis of these fluxes, Fluxite being one of the commercially available products.

For small electrical joints, **CORED SOLDER**, where the flux is carried in a hollow wire of solder so that it can be applied simultaneously with the solder, is very convenient.

For most work, a soldering iron is used: "iron" is a misnomer as the "bit," the business end of the tool, is made of copper. This metal is used because it is a very good conductor of heat and rapidly transfers

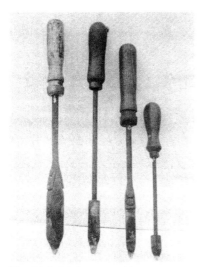

Fig. 14.3 *Solid copper soldering irons.*

heat from the iron itself to the work, and it readily alloys with tin. This latter property allows the tip of the iron to be "tinned," that is given a coating of solder. A selection of soldering irons appears in Fig. 14.3.

The most convenient way of heating an iron is on a gas heater. In industry, a rectangular gas heater, covered by an iron box, is used, the copper bit being placed in the box in the flame. If this special facility is not available, any form of gas "ring" will do. I have no gas supply in my workshop and I use a paraffin blow lamp with a simple stand to carry the irons as shown in Fig. 14.4.

Electric irons, as shown in Fig. 14.5, are available and are useful for small work, particularly for soldering electric cables. For heavier work, I find the ordinary iron, heated as described above, far superior. It is an advantage to have two irons so that one can be heated while the other is being used.

The first operation is to "tin" the iron. This is done by heating it to a high temperature, but less than red heat. The tip is then quickly cleaned with an old file and then dipped in the flux. If the temperature and other conditions are right when the tip is rubbed with solder, it will take on a thin film of solder. I find an easy way of doing

Fig. 14.4 *Paraffin blow lamp with stand for heating soldering irons.*

this is to pour a small drop of Baker's Fluid onto a tin; an old biscuit tin lid is ideal. If the hot and clean iron is then rubbed in the flux and solder applied, a lovely bright silver-colored surface appears on the tip. When in this condition, it must not be overheated, otherwise the tinning process will have to be repeated.

When making a lap joint, the minimum overlap is three times the thickness of the metal. The optimum solder film is 0.003 in. See Fig. 14.6. Both the surfaces to be joined must be cleaned and tinned. Assuming Baker's Fluid is being used, this can be done by dipping the solder stick in the flux and then applying the solder to the iron, flux being added to the surface being soldered as required. The iron must be held in close contact with the work for long enough for the heat to be transferred to the work, as it is only when the work is really hot that the solder will run freely.

When both surfaces have been tinned, they are placed together after adding a little flux. The iron is then moved over the joint, causing the tinning to melt and to fuse the two materials together. A little extra solder can then be run along the

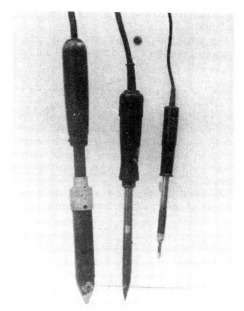

Fig. 14.5 *Electric soldering irons.*

edges of the joint to give added strength and a neat appearance.

A thick layer of solder between the two metals being joined does not make the joint as strong as when a small amount is used. As stated earlier, the optimum thickness of solder in the joint is 0.003 in.

As large an iron as can be handled should be used and it should be as hot as

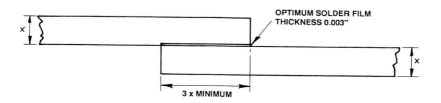

OPTIMUM SOLDER FILM
THICKNESS 0.003"

3 x MINIMUM

Fig. 14.6 *Lap joint minimum dimensions.*

possible but, of course, not red hot. If you doubt your ability to make a good joint, try yourself out on two pieces of scrap material and then, having made the joint, pull it apart to establish how sound a job you have made. All workshop skills can only be gained by practice; in this case make sure of clean work and a hot iron when you "have a go."

A word of warning concerning the repair of tanks that have contained a flammable liquid such as gasoline. Every care must be taken to ensure every trace of the liquid and its vapors are expelled from the tank before any heat is applied. Ideally, the tank should be steam cleaned, but if this is not possible, it should be filled with water, shaken, and then "topped" up with more water. It should then be shaken again and again "topped up." This process should be repeated until no more water can be added and it is certain that no pockets of flammable vapor remain. This must be done before any heat is applied, even if the tank is an old one from which it appears all the fluid has evaporated.

Chapter 15

Silver Soldering, Brazing, Bronze Welding, and Engineering Adhesives

SILVER SOLDERING is similar to soft soldering, except that the operation is carried out at a much higher temperature and the resulting joint is much stronger.

Silver solder is an alloy and alloys seldom have an exact melting point; because of this, silver solders are described as having a melting range. The solder will be solid at a temperature below the lowest figure quoted, the solidus, and liquid above the higher temperature, the liquidus.

Easy-flo silver solder, complying with B.S. 1845 AG1, supplied by Johnson Matthey Metals Ltd., is made up of 50% silver, 15% copper, 16% zinc, and 19% cadmium. It has a melting range of $1148°–1166°F$ $(620°–630°C)$.

Easy-flo silver solder No 2, complying with BS 1845 AG2, contains 42% silver, 17% copper, 16% zinc, and 25% cadmium and has a melting range $1126°–1143°F$ $(608°–617°C)$.

Easy-flo contains more silver than Easy-flo No 2 and, consequently, costs more. Its higher silver content gives Easy-flo somewhat higher ductility in the cast condition, better corrosion resistance, and it provides neater joint fillets than can be obtained with Easy-flo No 2. There is no practical difference in the melting temperatures, or in the melting ranges of the two alloys. Easy-flo No 2 is recommended in all cases, excepting those where maximum ductility is essential, where the smoothest fillets are desired, or where the corrosion conditions of service are severe.

FLUXES. Because of their low melting temperatures, Easy-flo solders should not generally be used with relatively high-melting point brazing fluxes based on borax or borax acid. For the best results, use should be made of Easy-flo flux, except for special cases.

It is important that a gap is left in the joint into which the solder can run by capillary action. There MUST be a gap for the solder to penetrate if a proper joint is to be made. A gap of 0.001 in. (0.025mm) is the minimum required and double this figure is more likely to be satisfactory.

Cleanliness of the parts to be joined is essential in both brazing and silver soldering. The surfaces to be joined must be rendered free from dirt, rust, and grease by filing, scraping, or grinding.

Where a paste flux is used, and this type is recommended, it should be smeared on the joint before placing the two pieces together. If the flux is in powdered form, a small quantity should be mixed into a paste and used in the same way.

The actual way in which silver soldering is tackled will depend on the type of heat available and the size of the job to be tackled. Some form of torch is essential. The oxy-acetylene welding outfit, described in the next chapter, can be used for this purpose. Excellent results can be obtained with the Primus-Sievert L.P.G. appliances or even with the humble propane torch.

Clamps may be necessary to hold the pieces to be joined in position while the soldering is carried out. Toolmaker's clamps, C-clamps, or wire may be used for this purpose. Tubal Cain, in his admirable book, *Simple Workshop Devices,* describes an easily made little gadget for holding delicate pieces of work in position.

Having rendered the joint scrupulously clean and smeared it with paste flux, the work is heated until it is hot enough to melt the solder, which will then run into the joint.

Pickling, that is immersing in acid, is not necessary for the removal of most brazing fluxes. The work should be quenched from black heat in warm water, when most of the flux will crack or dissolve. Sulfuric acid pickle will remove oxide or scale and hasten flux removal. The best proportion for the pickle is one part of concentrated sulfuric acid to ten parts of water. Old battery acid is not recommended, as it may contain lead sulfate, which may react on the work. When making this solution, the acid must be carefully and slowly poured into the water and not water into the acid. This is because the mixing of the two liquids produces considerable heat. If the acid is added to the water, the mass of water absorbs the heat. If the water is added to the acid, the heat produced will cause the acid and water to be thrown out of the vessels with such violence that the person doing this mixing is virtually certain to receive acid burns. Diluted acid can be obtained from some chemists and it is better to purchase it in this way than have the trouble and risk of mixing it. Even diluted acid is dangerous—it will burn holes in clothes and it must be used with great care.

BRAZING, a similar process to silver soldering, is where metal parts are joined by flowing melted brass between the surfaces to be joined. A much higher temperature than that required for silver soldering is necessary, a red heat as high as 1562°F (850°C) being required. The process cannot be used, of course, where the metal to be joined has a lower melting point than the brass.

BORAX is the flux generally used; a small quantity is made into a paste with water and applied to the parts to be brazed before heating. The rest is kept dry for use during the operation.

BRONZE WELDING. Although this term is in general use, it is something of a misnomer. It is not true fusion welding because the metals joined are not heated to melting point. It cannot be correctly described as brazing because brazing involves capillary attraction of the filler metal into a narrow joint.

Bronze welding involves the use of alloy bronze rods and is used for making joints in copper, for joining dissimilar metals, and for repairs to cast iron. It is essential that the edges of the materials being joined are not melted but merely raised to red heat. For example, when bronze-welding cast iron, Saffire manganese bronze rod should be used, melting at 1634°F (890°C), well below the melting point of cast iron.

ENGINEERING ADHESIVES

In recent years, a very large range of adhesives has been made available to the engineer, and many of them have been put to uses that the older engineer never thought possible. For example, it is possible to use an adhesive to retain a component in position instead of using an interference fit. This has the advantage of eliminating the stress that an interference fit causes and allows machining tolerances and surface finishes to be eased. This method has been used to retain all the parts of an aluminum racing cycle frame and the main column of a milling machine.

EPOXY ADHESIVES. These come in two tubes, one containing an adhesive and the other a hardener. When mixed together, in equal quantities, a chemical reaction occurs between the two, which results in a very hard and strong bonding between the joint faces to which they are applied. Araldite is probably the most famous of this type and it is supplied in two kinds. The original type takes twenty four hours to cure at a temperature of 68°F (20°C), but can be handled after three to five hours if kept at room temperature. More recently, a "rapid" version has become available, which has a much quicker initial set. This type has the disadvantage of a very short life as a usable "mix," so it has to be applied very quickly. Joint faces must be clean and grease-free.

ANAEROBIC ADHESIVES. These adhesives have the peculiar property of remaining fluid until air is removed from them. As soon as the fluid is smeared onto two surfaces and the surfaces are brought together, the fluid begins to harden or cure. The components to be joined must be free from all traces of dirt and grease. This cleaning can be done using a rag soaked in white spirit or gasoline (take the usual precautions against fire). Washing with a detergent and water is also helpful. A good grip is obtained in about two hours but full strength is not obtained for about twelve hours.

Special types are available for threadlocking, making gaskets, sealing pipe connections, and for bonding a variety of materials.

CYANO-ACRYLATE ADHESIVES are very fast curing compounds, marketed by one company under the name "super-glue." They set within seconds. Care is needed in their use as drops spilled on human skin bond almost instantly. If this occurs, for example, if the fingers get stuck together, they may be carefully peeled apart. Deposits on the skin should be washed in warm soapy water and the deposit peeled off. It is advisable to wear goggles when using this type of adhesive.

As with the other types of adhesives, cleanliness of the parts to be joined is important; they should be washed in a detergent and water.

There are many different adhesives available, each excellent for the particular purpose for which they are designed, and it is vitally important to study the maker's specification charts to ensure that the type most suitable for the job in hand is used. Loctite and Permabond are two companies specializing in this field.

Chapter 16

Welding

The art of welding has been known and practiced for many centuries, but until relatively recent times, the only practical method of making a weld was by heating the two pieces to be joined at the point where the joint was to be made, then, when they were almost at fusing point, placing them together in their proper relationship on an anvil and completing the weld by hammering. This process is known as **FORGE WELDING**.

This technique is unlikely to be available to the amateur because, generally, they have insufficient heat available to bring the metal to the required temperature. Blacksmiths are the experts in this type of welding and a forge, similar to the one they use, is almost essential. Before looking at the more common forms of welding, a brief look will be taken at what this older type of welding has to offer.

Wrought iron and mild steel are the metals that can be welded in this way. The welding heat for iron is at a temperature of about 2462°F (1350°C), when the metal is white hot and bordering on the pasty stage. Mild steel is welded at a slightly lower temperature when its color is yellow and before merging to white. It is important that welding is done at the right temperature, because if it is too low no amount of hammering will cause the weld to take place, while if the temperature is too high the metal will be burned.

The scarf weld, Fig. 16.1a, is the most straightforward to carry out and the most often used. After the ends of the joint have been prepared, they are heated to

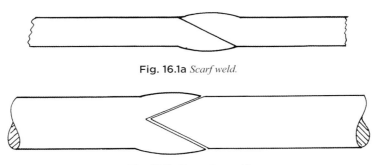

Fig. 16.1a *Scarf weld.*

Fig. 16.1b *V or splice weld.*

welding heat and then hammered together to complete the weld.

The "V" or splice weld is a very strong job and is preferred where the thickness of the metal permits preparation of the "V." One of the pieces to be joined is thickened up and then split with a hot chisel to form the "V." The other piece is forged to the shape shown in Fig. 16.1b, which shows the two pieces ready for welding.

OXY-ACETYLENE WELDING

If suitable proportions of oxygen and acetylene gases are mixed together and fed through the nozzle of a torch, the combined gases, when ignited, rapidly generate intense heat. The heat can quickly bring metals up to temperatures at which they can be successfully joined—or welded—together.

The acetylene and oxygen are stored in high-pressure cylinders fitted with regulators to reduce the high pressure to the considerably lower pressure required for the operation of the torch.

Murex Welding Products Ltd., ESAB Group (UK) Ltd, supply a lightweight portable oxy-acetylene welding set, known as the Portapak, a photo of which appears in Fig. 16.2.

Acetylene cylinders are painted maroon and contain a porous substance and a solvent for the gas. They are charged at pressures corresponding to 225 lbs/in^2. Because acetylene cylinders contain liquids, *they should always be stored and carried upright.* All acetylene

Fig. 16.2 *Portapak welding set.*

fittings have left hand threads, so it is impossible to connect them, in error, to the oxygen cylinders. The acetylene hoses are red in color.

Acetylene is a highly flammable gas, and if allowed to mix with air, is likely to explode if ignited by flame, heat, or spark in the vicinity. See, therefore, that all joints, especially those at the cylinder valve that are under high pressure, are gas tight, hose in good condition, and gases turned off at the cylinders when work is finished. Do not test for leakage with a flame; use soapy water.

If an acetylene cylinder leaks at the gland around the valve spindle, the leakage can usually be corrected by tightening the gland nut. If the cylinder leaks at the valve or the base plug, and the leak cannot be remedied by firmly closing the

101

valve, the cylinder should be moved into the open away from fires, electric motors, and such sources of spark or heat. The suppliers should be advised immediately. Forbid smoking and naked lights near the leaking cylinder.

If acetylene from the cylinder catches fire at the valve or regulator due to leakage at the connection, shut the valve and make the joint properly tight before further use.

Action in the event of a fire in which cylinders are involved. Gas cylinders should always be treated carefully but experience has shown that they are in fact high integrity packages which withstand considerable abuse from users. If, however, they are subject to extreme heat in a fire, they can explode. The following is good advice to anyone faced with the situation of a fire in which gas cylinders are known to be involved.

Keep everyone well clear until the fire department arrives to take control.

Inform the fire department immediately of the location and type of any gas cylinders involved in the fire. Also, tell them the location and type of other gas cylinders in the premises.

Cylinders that are not involved in the fire, and which have not become heated, should be moved as quickly as possible to a safe place, providing this can be done without undue risk. Make sure cylinder valves are closed.

Cylinders in the fire should be cooled by spraying with copious quantities of water over the entire exposed surface. Personnel engaged in this should take up positions that will give protection from exploding cylinders.

Great caution should be taken after the fire has been extinguished as there is still the possible danger that cylinders affected by heat could explode.

(Action in the event of a fire reproduced by kind permission of B.O.C. Ltd).

Oxygen cylinders are painted black and are fitted with green hoses. There is no porous mass or liquid in them. NEVER, UNDER ANY CIRCUMSTANCES, allow oil or grease to come in contact with oxygen fittings. Oxygen escaping from a leaking hose will form a flammable mixture with oil or grease, and may cause clothing and other articles to take fire vigorously from a spark.

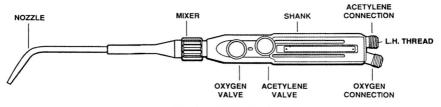

Fig. 16.3 *Welding torch.*

The torch for the Portapak is shown in Fig. 16.3. A key is provided to open the cylinder valves and they should be opened slowly. Sudden, rapid opening may cause damage to the regulators.

After consulting the data chart provided, the oxygen regulator is adjusted to the required working pressure. This must be done with the green oxygen valve on the blowpipe open. This valve is then closed and the acetylene working pressure is adjusted in the same way.

A small quantity of gas is then fed through each hose to clean the passages. This must be done one at a time, ensuring one valve is closed before opening the other one.

A spark gun is supplied with the outfit, and after opening the red acetylene valve slightly, the spark lighter is held facing along the direction of the flame at the end of the nozzle and the gas is ignited (see Fig. 16.4). The flame is adjusted by opening the red acetylene valve until the

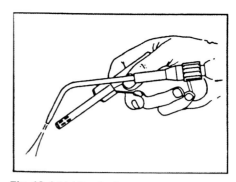

Fig. 16.4 *Lighting the flame.*

flame ceases to smoke. The green oxygen valve is then gently opened.

Three types of flame can be obtained. For welding steel, cast iron, copper, and aluminum, a neutral flame is necessary; this is where equal amounts of oxygen and acetylene are being burned at the same rate, and will be evident by the white cone being clearly defined with the merest trace of acetylene haze (see Fig. 16.5b).

For bronze welding, an oxidizing flame is required and this is obtained by

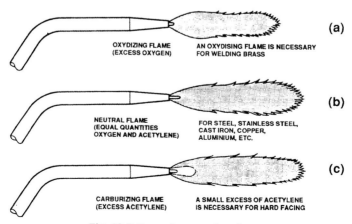

(a)

OXYDIZING FLAME
(EXCESS OXYGEN)

AN OXYDISING FLAME IS NECESSARY
FOR WELDING BRASS

(b)

NEUTRAL FLAME
(EQUAL QUANTITIES
OXYGEN AND ACETYLENE)

FOR STEEL, STAINLESS STEEL,
CAST IRON, COPPER,
ALUMINIUM, ETC.

(c)

CARBURIZING FLAME
(EXCESS ACETYLENE)

A SMALL EXCESS OF ACETYLENE
IS NECESSARY FOR HARD FACING

Fig. 16.5 *Types of oxy-acetylene flames.*

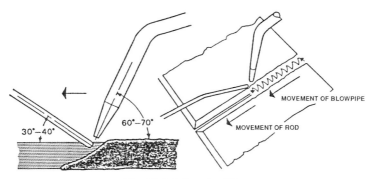

Fig. 16.6 *Leftward welding.*

increasing the amount of oxygen. The flame then appears as is shown in Fig. 16.5a. For hard facing and certain other specialized applications, a carburizing flame is required and this is obtained by increasing the flow of acetylene until a haze or feather of acetylene is evident at the end of the white cone (see Fig. 16.5c).

LEFTWARD WELDING. Welding is a specialized subject and it is not proposed to discuss the various techniques in great detail. Leftward welding is the most commonly used method and is used on steel for flanged edge welds, for unbeveled plates up to 3.2mm (⅛ in.) and for beveled plates up to 4.5mm (³⁄₁₆ in.). It is also the method usually adopted for cast iron and non-ferrous metals.

Welding is started at the right-hand edge of the joint and proceeds toward the left. The torch is given a forward motion with a slight sideways movement to maintain melting of the edges of both plates at the desired rate, and the welding rod is moved progressively along the weld seam (see Fig. 16.6).

This form of welding differs from that used by the blacksmith in that the edges of the metal being joined are actually made molten and fuse together. Any lack of metal in the joint is made up by a filler rod, of the appropriate material, which is melted in the seam as the weld progresses.

RIGHTWARD WELDING procedure is recommended for steel plate over 5.00mm (³⁄₁₆ in.) thick. Plates from 5.00mm to 8.00mm (³⁄₁₆ in. to ⁵⁄₁₆ in.) need not be beveled. The weld is started at the left of the seam and proceeds toward the right as shown in Fig. 16.7. The torch flame precedes the filler rod in the direction of travel. The rod is given a circular forward motion and the torch is moved steadily along the weld seam.

Special filler rods and fluxes are required for metals other than mild steel. For all but the smallest cast iron articles, it is necessary to pre-heat the metal before welding and to allow very gradual cooling after the weld has been completed, to allow for the effects of expansion and contraction as the temperature varies.

104

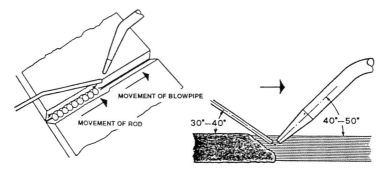

Fig. 16.7 *Rightward welding.*

Goggles, supplied with the equipment, must always be worn to protect the eyes from dangerous sparks and rays. Quite apart from the safety aspect, the goggles enable the workpiece to be seen more clearly.

METAL CUTTING PROCESS USING OXY-ACETYLENE GAS

It is possible to cut steel with an oxy-acetylene outfit by changing the welding torch for the cutting one shown in Fig. 16.8. The chief difference in the two is that on the cutting torch an inner nozzle carries a supply of pure oxygen that is used for the actual cutting or burning of the metal. The outer nozzle supplies a mixture of oxygen and acetylene gases, which ignite and are used to preheat the work. When a jet of oxygen is directed onto mild steel or wrought iron that has been preheated to red heat, the metal ignites, thereby providing its own fuel, burns away rapidly, and forms a deposit of iron oxide. The pressure from the jet blows away this deposit, and leaves a clear-cut line in the metal as the torch is moved along. This is because the iron oxide thus formed melts at a lower temperature than mild steel or wrought

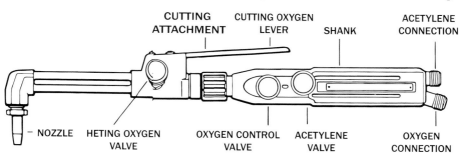

Fig. 16.8 *Cutting torch.*

105

iron, and due to the intense heat generated by the oxygen, the oxide, in liquid form, is easily blown away out of the path of the cut.

The blowpipe is held at right angles to the work to be cut and the flame applied to the edge farthest away from the operator. When the metal reaches a red heat, the cutting oxygen stream is released and the blowpipe moved slowly along the line to be cut.

ARC WELDING

The art of welding metals together by means of electricity is not a new process. American Professor Elihu Thomson, the father of the art, patented the first practical electric welding machine in 1886.

An electric arc is simply a sustained spark between two terminals. In arc welding, the spark is formed between an electrode, connected to the welding machine and held in the operator's hand, and the work being welded. When the electrode is scraped on the job, sparking occurs, due to the imperfect electrical contact being made. The heat from these sparks ionizes the air in their immediate vicinity and makes it conductive to electricity. If the electrode is now withdrawn slightly, the electric current continues to flow across the gap, through the ionized air, forming an arc.

A coated electrode is used. The heat of the arc burns the electrode coating, which gives off inert gases. The gases surround the arc and prevent contact between the weld metal and the oxygen and nitrogen of the air (Fig. 16.9). The coating is consumed at a slower rate than the core wire of the electrode, and this causes the end of the coating to project beyond the end of the core wire, thus helping the operator to direct and control his arc. The coating leaves a slight slag on top of the weld, which protects the weld metal while cooling. This slag is chipped away when the work has cooled.

Most workshops will have an alternating current supply and, therefore, a transformer-type welding set connected to this supply, with means of adjusting the output to the needs of different jobs, will be used.

To protect the operator's face and eyes from the direct rays of the arc, it is essential that a face mask should be used. These are generally constructed of some kind of pressed-fiber insulating material, dead black in color to reduce reflection. The shield is fitted with a glass window of such composition as to absorb the infra-red rays, the ultraviolet rays, and most visible rays emanating from the arc.

INERT GAS SHIELDED WELDING

This type of welding has become popular because much higher welding speeds are obtainable than with normal gas or arc welding; consequently, there is less heat input to the material being welded and less distortion, which results in production costs being reduced. Although this process

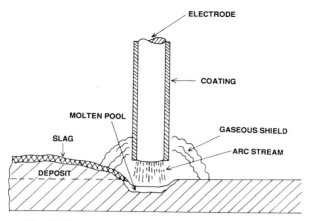

Fig. 16.9 *How the coated electrode provides the gas shield.*

has traditionally been for the professional (because of the expensive equipment needed), the prices of equipment are coming down, and a brief description of the techniques involved should be of interest to all model engineers.

The two main types have become known as TIG and MIG welding, the letters standing for Tungsten-Inert-Gas and Metal-Inert-Gas respectively. These are fusion welding processes where the weld pool and the arc are enveloped by an inert gas, delivered to the weld point through the welding torch or gun. The molten metal is completely shielded from atmospheric contamination. Argon is the gas most commonly used in Britain.

With TIG welding, the electrode is not consumed and is not used as a filler metal. Where a filler metal is required, a separate welding wire is applied in much the same way as in oxy-acetylene welding.

In the MIG welding process, a small wire is drawn from a reel by electrically driven rollers, into the welding gun held by the welder. This fine wire, between ⅟₃₂ in. and ³⁄₃₂ in., and of the required composition, is fed at a constant speed through the gun, to an arc maintained between the wire and the work.

During most of the welding processes, some sparks and globules of molten metal are thrown out and protection from possible burns is prudent. A leather apron is ideal and stout footwear is desirable.

As stated earlier, welding is a specialized subject, which is impossible to cover fully in a book of this kind. As is the case with most workshop topics, practical experience is very necessary. Technical colleges and suppliers of welding equipment run courses on welding techniques and these are highly recommended.

Chapter 17

Hardening and Tempering Tools

The first step in making a hardened steel tool is to make it dead hard. In this condition, it is too brittle to be used and the hardness must be let down to a degree suitable for the job the tool is intended to do. The latter process is known as **TEMPERING**. There is no need to go into the changes that take place in the structure of the metal during this hardening and tempering process; it is sufficient to say that when carbon steel is heated to about 1346°F (730°C) and kept at that heat for a short period of time, it changes to a hardened state. If it is allowed to cool in the ordinary way, it will return to its soft condition. If, however, it is cooled quickly, it will remain hard.

As was said earlier, it is impossible to harden mild steel by heating and quenching because of its low carbon content, but as the carbon content of the steel increases, so the harder it becomes when heated and rapidly cooled, as the accompanying table shows.

Mild steel can be given a hard outer skin by a treatment known as case-hardening, which will be described later.

The tool to be hardened should be heated up slowly at first, bringing it over the initial, or black hot, stage comparatively slowly so that the steel can adapt itself to the change of form. Once it begins to show signs of changing color, it may be heated more vigorously. Sometimes cracks appear in a tool that has been hardened, and this initial careful heating and care taken later not to overheat the steel will go some way in preventing this.

When a forge is used to heat the steel, the tool should be turned over and around in order to present all the sides to the heating source, so that the heating is constant over that part of the tool that has to be hardened.

Tools such as chisels and lathe tools only need to be hardened at their cutting

TYPE OF STEEL	CARBON CONTENT	EFFECT OF HEATING AND QUENCHING
Mild	Below 0.25%	Negligible
Medium carbon	0.3 to 0.5%	Becomes tougher
Medium carbon	0.5 to 0.9%	Becomes hard
High carbon	0.9 to 1.3%	Becomes very hard

edge and with a small margin to allow for sharpening by grinding. When heating this type of tool, arrange for that part of the tool where the change from hard to soft takes place to be heated, without a sharp demarcation between the red and black hot parts, so that when the tool is tempered, there is a gradual change from the hard to the soft portion.

The actual temperature at which the steel should be quenched depends upon its composition. For example, drill rod should be brought to a temperature of 1418°F to 1454°F (770°–790°C), while a chromium high carbon alloy steel is hardened at 1742°F (950°C) to 1832°F (1000°C). In most general workshops there is no means of accurately measuring temperatures, so it is a question of judging the temperature by the color of the heated steel. The hardening heat corresponds to a bright cherry red when viewed in daylight . The tool should be kept at this temperature for about half a minute before quenching to allow a thorough transformation to occur.

Another way of judging the correct temperature for hardening is by using a magnet. As the metal is heated, it reaches a temperature where it is no longer attracted by a magnet. Test by just touching the tool with the magnet, do not let the magnet get hot or it will be demagnetized. A temperature 140°F (60°C) above the temperature at which this phenomenon occurs will generally be about right for hardening.

Care must be taken when dealing with tools with sharp cutting edges or points not to direct the hot portion of the flame directly at the edge or point for any length of time. Because of their small volume, it is very easy to overheat and burn them, so that after hardening and tempering, they are very brittle. No amount of re-hardening and tempering will remedy this condition and the defective point or edge must be ground away and a fresh start made.

It is a good idea to keep an old file to find out if the hardening has been successful. If the file will bite into the metal, it is certainly not hard and a further attempt at hardening must be made.

Most ordinary carbon steel has a mottled gray appearance over all the dead hard parts. This mottling takes the form of almost white or bright patches; the clearer and closer together they are, the harder the steel will be.

TEMPERING

In an ideal situation, tools to be hardened are heated in a special furnace that is maintained at a suitable constant temperature as indicated on a high reading thermometer or pyrometer. When tools are made singly or in small quantities, as applies in most workshops, they are heated by a Bunsen burner, propane torch, gas torch, or a small blacksmith's forge, with no means of accurately controlling the temperature. However, this is not an insurmountable difficulty.

COLOR	TEMPERATURE DEGREES C	TOOLS
Pale yellow	428°F (220°C)	Hand turning tools and hammer faces
Pale straw	446°F (230°C)	Turning tools
Dark straw	464°F (240°C)	Drills and milling cutters
Brown	482°F (250°C)	Taps
Brownish purple	500°F (260°C)	Punches, reamers, twist drills, and rivet snaps
Purple	536°F (280°C)	Cold chisels
Blue	572°F (300°C)	Springs

All steels, hardened or not, that have a bright surface, free from moisture, become coated with an increasing thickness of oxide when heated. This oxide presents varying shades of color that range from a very pale yellow to a brilliant dark blue. The table above gives the temperatures at which the various colors appear and the tools that can be tempered at each temperature.

It is often not necessary to harden the whole of a tool but only that portion that is required to do the cutting. If this is the case, then only that portion required to be hard should be heated and quenched. This leaves a hard cutting portion with a softer but tough shank. Tools such as chisels and punches must not be hardened at the end where they are struck with the hammer. If that end is hard, it will be brittle and chips are likely to break off and fly into the face or hands of the operator or anyone standing nearby.

The portion of the tool to be tempered should be polished with emery cloth until it is bright so the tempering colors can be clearly seen. When the appropriate color appears, the tool is quenched.

Clean water is an effective coolant but it can be improved by adding salt, ¾lb to a gallon of water. The tool should be immersed in the coolant vertically and point downwards and with a moderate up and down movement. It must not be quenched in a horizontal position as it may contract more on one side than the other, causing cracking.

For some tools, such as chisels and punches, the one-heat method of heating and quenching, used by blacksmiths, may be used. The tool is heated to the hardening temperature for about half its length and then the cutting edge is quenched for a length of about 30 to 60mm. When it is quite certain that this quenching is complete, the still-hot tool is removed from the water and the cutting edge polished with a piece of a broken emery wheel, or if this is not available, with a strip of emery cloth nailed to a piece of wood. The heat from the unquenched portion will travel, by conduction, to the cutting edge and the tempering colors will appear. When the required color appears, the whole tool must be quenched. This method gives a tool hardened at its cutting edge with the metal gradually decreasing in hardness toward the shank, leaving the end soft where it is struck by the hammer.

Like all workshop jobs, actual experience is far better than the written word and beginners are advised to practice until they master the technique. Unlike most workshop jobs, if a mistake is made in hardening and tempering, the tool is not ruined. It can be re-hardened and tempered to remedy any error that was made at the first attempt.

CASE HARDENING

Case-hardening is employed on iron and mild steel that cannot be hardened by heating and quenching because of its low carbon content. It does not take the place of real hardening because only the surface is hardened for a small depth; in other words, it imparts a hard skin to the steel, while the center remains soft. Obviously it is not a suitable medium for a cutting tool that has to be sharpened, because the case-hardening would be ground away. Its main

use is to superficially harden the wearing surfaces of such items as pins and journals so that they resist wear but at the same time retain the greater part of their toughness.

Ideally, the objects to be case-hardened are packed into cast iron or steel boxes and covered all over with a substance rich in carbon, such as charcoal granules, hoof clippings, bone dust, or charred leather. The boxes and their contents are then placed in a furnace at a temperature of about 1652° to 1742°F (900°–950°C) for some hours. Following the heating, the boxes are opened while hot and the steel pieces dipped into the quenching bath, usually containing cold water.

It is unlikely that the amateur will be able to use this method but good results can be obtained by using one of the trade products. "Cherry Red" is a very well known one sold in powder form.

To case-harden small parts, some of the powder is heaped on to a metal tray. The job is heated to a yellow heat and then rolled in the powder. This is repeated two or three times, and then the component is again heated and finally quenched.

Chapter 18

Keys, Keyways, Splines, Collars, and Shafts

In engineering parlance, a **KEY** is a component used to locate and secure an object in position. In its most common form, it is used to position a pulley or wheel on a shaft and to ensure that they rotate together at the same speed.

In addition to securing a wheel to a shaft, the key also has to take the shear stress when the wheel is turned, and the turning effort of the wheel has to be transmitted to the shaft. In other cases, the shaft is driven and the key has to take the drive to the pulley. Because of this shear stress, the key must be a good fit and bear evenly on all faces of the wheel boss keyway (that is the recess in which the key fits) and in the shaft keyway. If the key is a poor fit, the shear stress is transmitted only by the parts that are in close contact instead of the whole area. The wheel will soon begin to rock on the shaft and rapid wear will occur both on the key and the keyway.

The **GIB HEAD KEY**, illustrated in Fig. 18.1, is rectangular in section and the top face has a taper of approximately 1 in 64, but its sides are parallel.

When a key of this type has to be removed, a wedge is driven between the head of the key and the wheel hub (see Fig. 18.2a). On the rare occasions, the back end of the key is accessible and a drift may be used to drive the key out as is shown in Fig. 18.2b.

When fitting a key, its sharp corners should be filed off, as these give a false impression that the key is fitting tightly. The key is then filed so that it is a nice sliding fit, without shake, in the keyways.

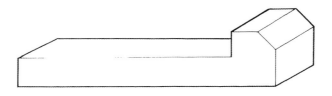

Fig. 18.1 *Gib Head key.*

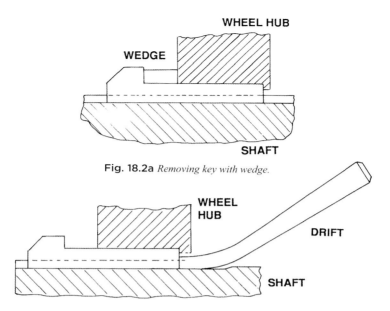

Fig. 18.2a *Removing key with wedge.*

Fig. 18.2b *Removing key with drift.*

The point of the key is then filed down for a short distance until it will just enter the two keyways when the wheel is on the shaft. This gives a rough indication of the depth of the key. The filed-down portion should then be carefully measured with a micrometer. The top of the key is then filed down until the part near the head is a few thousandths wider than the micrometer reading. The sharp corners are again removed and the key tried in the position it will eventually occupy. It is gently driven in position, removed, and the top inspected for friction marks, which will show clearly where it is touching the top of the wheel keyway. These friction spots are carefully removed with a scraper and the key again inserted in position.

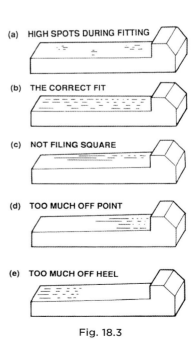

(a) HIGH SPOTS DURING FITTING

(b) THE CORRECT FIT

(c) NOT FILING SQUARE

(d) TOO MUCH OFF POINT

(e) TOO MUCH OFF HEEL

Fig. 18.3

113

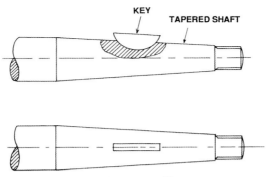

Fig. 18.4 *Woodruff key.*

This process is repeated until the whole area of the top of the key shows compression streaks covering practically the whole surface. Fig. 18.3a shows the key with high spots during fitting, 18.3b the ideal fit, and 18.3c a fault produced by not filing square across the top of the key. Filing too much off the point of the key or at the opposite end will give the marks depicted in Figs.18.3d and 18.3e. Marking blue may be used to make the marks where the key is bearing more easily seen.

THE WOODRUFF KEY is completely different from the Gib Head key already described. Its side view is the minor segment of a circle but it is of even width and thickness. The curved part fits into a similarly shaped recess in the shaft. It is used chiefly on tapered shafts as shown in Fig. 18.4. The hub of the wheel has a similar taper to the shaft. The shaft end is usually threaded and fitted with a nut and washer, so that when the nut is tightened, the boss grips the shaft firmly with a wedging action so that the key is relieved of shear stress. There must always be clearance between the top of the key and the keyway in the wheel, since if this were not so, the key would prevent the hub being tightly wedged on to the shaft.

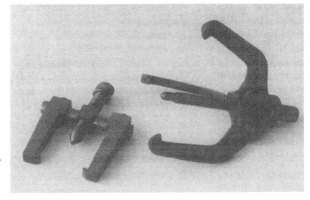

Fig. 18.5 *Two and three leg wheelpullers.*

114

The keyway on the shaft is machined with a special cutter in a milling machine, a process which is outside the scope of this book, but benchworkers may be required to make the key. The key is cut from a piece of round steel of a diameter to match the curve of the key, and then filed so that it is a nice fit in the keyways, remembering that clearance must be maintained between the top of the key and the keyway in the wheel.

The removal of a wheel fitted to a tapered shaft is often difficult, especially if it has been in position for a long time. Generally, the wheel is so firmly fixed that it can only be removed with the aid of a wheel-puller. These are of many different types: one with three legs and another with two are shown in Fig. 18.5. The pull exerted by the legs of the puller should be as near the hub of the wheel as possible. Pulling on the rim, as shown in Fig. 18.6, often cannot be avoided, but this puts a big strain on the wheel and makes the puller less effective than if the legs were nearer

the hub. In some cases, tapped holes are provided in the wheel for the attachment of the puller. When tightening, if the screw of the puller does not remove the wheel, a smart hammer blow to the head of the screw will often have the desired result. The sudden jar of the hammer blow causes the tightly wedged taper to break away, while the steady pull of the wheel-puller fails to do so.

Woodruff keys are sometimes used on parallel shafts where the load is light, but in that situation, the key has to resist shear stress as there is no wedging action between the shaft and the wheel.

A FEATHER KEY, Fig. 18.7, is of rectangular section, usually with rounded ends. It is sunk for about half its depth into the shaft and its sides and ends fit snugly all around the keyway. It is used where clutches, gears, and other details are required to slide along a shaft. It is often necessary to allow this sliding movement in order to permit a wheel to

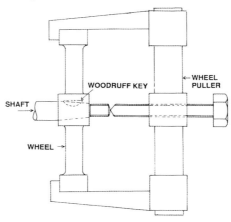

Fig. 18.6 *Removing wheel with wheel-puller.*

115

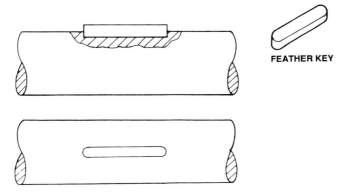

FEATHER KEY

Fig. 18.7 *Feather key.*

engage with another gear-wheel parallel to it. The keyway on the shaft will have been machined on a milling machine, but the key may have to be made at the bench. The wheel must be free to slide on the shaft, so the key must, therefore, be a sliding fit in the wheel keyway and a snug fit in the shaft, and there must be clearance between the top of the key and the hub keyway. Sometimes this type of key is held in the shaft by countersunk screws.

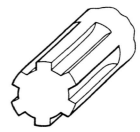

Fig. 18.8 *Splined shaft.*

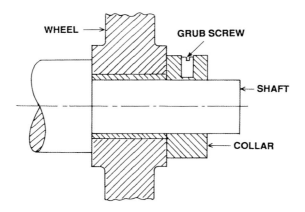

WHEEL

GRUB SCREW

SHAFT

COLLAR

Fig. 18.9 *Collar fitted to prevent wheel moving endways on shaft.*

SPLINED SHAFTS provide a similar sideways movement for a wheel carried on them to that described for a feather key. The shaft has rectangular recesses all around its circumference and the wheel hub is similarly shaped. Splines are used extensively in all types of gearboxes where the various gear ratios are obtained by sliding gearwheels along shafts to mesh with other wheels (See Fig. 18.8).

In some cases, "V" shaped splines are used to prevent a wheel from turning on a shaft. Steering wheels on motor cars, for example, are generally fitted to the steering column in this way.

To prevent a wheel fitted freely on a shaft from moving sideways, a **COLLAR**, Fig. 18.9 is used. It consists of a plain ring fitted with a grub screw (set screw) to hold it firmly on the shaft.

Chapter 19

Sheet Metalwork

Sheet metalwork is the province of skilled tradesmen—the silversmith, tinsmith, and panel beater—but on occasions the general engineer will be called upon to make articles using sheet metal. This is particularly the case with model engineers; locomotive cabs, tenders, and boilers are all fashioned from sheet brass and copper.

In the tinsmith's shop, many special tools and machines are employed. It is unlikely that these will be available to the model engineer or in the fitting shop, but much can be achieved by improvisation and with simple tools.

The metals used may be tinplate, black, bright or galvanized iron, copper, brass, or aluminum, or the precious metals, i.e., silver or gold, although it unlikely that the latter will be found in the ordinary shop.

The thickness of sheet metals may be given in **STANDARD WIRE GAUGE (SWG)** or in metric or imperial measure.

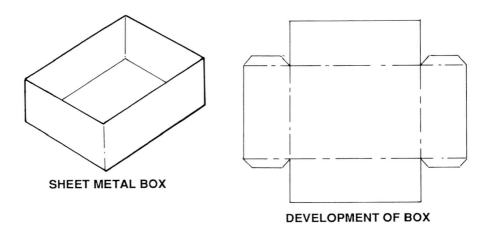

SHEET METAL BOX

DEVELOPMENT OF BOX

Fig. 19.1

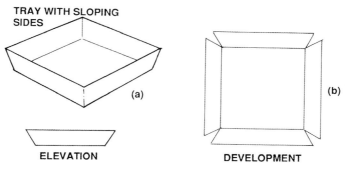

TRAY WITH SLOPING SIDES

(a)

(b)

ELEVATION

DEVELOPMENT

Fig. 19.2 *Development of tray with sloping sides.*

SWG	mm	Inch
22	0.7	0.028
20	0.9	0.033
18	1.2	0.048 (approx. $\frac{3}{64}$ in.)
16	1.6	0.064 (approx. $\frac{1}{16}$ in.)
13	2.5	0.092 (approx. $\frac{3}{32}$ in.)

The table above shows some of the corresponding sizes.

When a shape is being marked out on a thin metal, so that when cut out and folded up it forms a complete article, it is said that the surface of the object has been **DEVELOPED**. Fig. 19.1 shows the development of a sheet metal box. Flaps have been added to the development to enable the sides to be fastened together.

Often, the true shape of the surface is not shown directly on the drawing from which the development is to be made. Fig. 19.2a shows a tray with sloping sides, and the exact shapes of the sides are not shown on the drawing. It is imagined that the sides are swung down to the horizontal plane and then projected onto the bottom of the tray, as shown in Fig. 19.2b. Before actually cutting out the developed shape

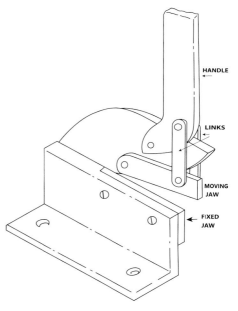

HANDLE

LINKS

MOVING JAW

FIXED JAW

Fig. 19.3 *Bench shears.*

119

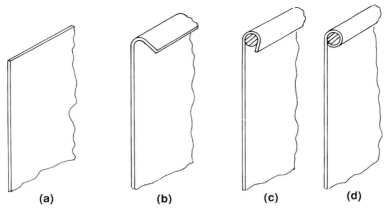

Fig. 19.4 *Stages in the formation of a wired edge.*

on the metal, it is often useful to do this on thin cardstock. This will show up any errors without spoiling valuable material.

Sheet metal is usually cut with bench shears, Fig. 19.3, or tinman's snips, which resemble very heavy scissors. They are made in a variety of forms, some specially made for cutting curves. In the tinsmith's shop foot-operated guillotines may be available and these machines give a very clean straight cut. Bench shears and guillotines are designed to slice through sheet steel with ease. Because of their power, they are also capable of inflicting serious injury if improperly used. Make sure you fully understand how to operate them and never use them without the appropriate guard in position.

ANNEALING is a heat treatment applied to metals in order to restore or increase their malleability. When a metal is worked from one shape to another,

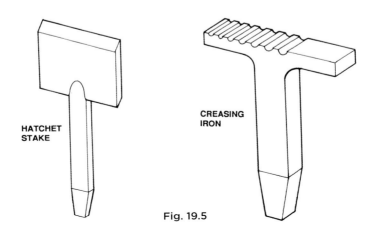

HATCHET STAKE

CREASING IRON

Fig. 19.5

whether it be by rolling, stretching, pressing, or flanging, it becomes harder under the process and is less malleable. In some cases, the metal will become so brittle that it will fracture if the annealing process is neglected. Copper is particularly prone to this work-hardening, as makers of model engine boilers will readily testify.

Generally, annealing is accomplished by heating the metal to a dull red and then allowing it to cool. Sometimes copper is plunged into cold water after heating and, although this is a time-saver and does not affect the annealing process, it may cause distortion. If there is any risk of this occurring, it is better to let the copper cool naturally.

Very often a sheet metal article has its edge bent over a wire so as to enclose and strengthen it. This conceals the raw sharp edge, which might prove a source of danger. Fig. 19.4 shows the various stages in the formation of an edge of this kind.

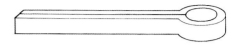

Fig. 19.6 *Folding iron forged from* ⁵⁄₁₆" x 1" *mild steel.*

The sheet-metal worker uses **STAKES** of various kinds, two of which are shown in Fig. 19.5. These are inserted in a special fixture on the bench. They are used, for example, to insert the wire in the edge of sheet metal, as illustrated in Fig. 19.4. Fig. 19.4a shows the elevation of the metal to be wired, (b) the wiring allowance after being bent over the hatchet stake by a mallet, and (c) the metal bent over the wire; in this operation the work would be supported on the creasing iron. Fig. 19.4d shows the wired edge with the

Fig. 19.7 *Folding iron held in vise to secure a piece of sheet metal in which a 90° bend is required.*

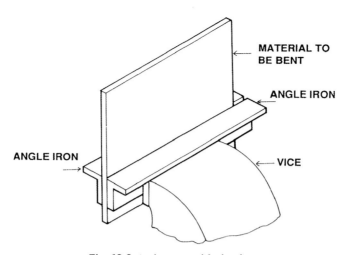

Fig. 19.8 *Angle iron used for bending.*

metal tucked neatly over the wire so as to enclose it completely.

The stakes are like miniature, specially shaped anvils, and if they are not available, much can be done by utilizing pieces of steel of similar shapes, held in the vise.

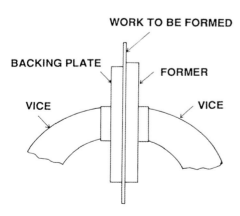

Fig. 19.9 *Set-up for forming work.*

A useful tool for holding sheet metal is the folding iron shown in Fig. 19.6, which is quite simple to make in the workshop if a forge or some other source of heat is available. Fig. 19.7 shows the folding iron held in the vise securing a piece of 20 SWG sheet metal which is to be bent over at a right angle.

Two pieces of angle iron, held in the vise, as shown in Fig. 19.8, can be used instead of a bending iron but this makes the job more difficult to line up.

BENDING ALLOWANCE. When sheet metal is bent, the outside of the bend becomes stretched and the inside compressed. Somewhere in the center of the bend there is a point where the metal is neither stretched or compressed. This is said to be about 0.4 times the thickness of the metal from the inside radius. If, therefore, a little under half the thickness of the metal is allowed for the bend, it should not be far out.

FLANGING AND FORMING.
Makers of model engine boilers are very familiar with this technique. Basically, it consists of making a **FORMER** over which the work is beaten into shape. Hardwood can be used for formers on which to flange copper up to about $\frac{5}{32}$ in. thick. Metal formers can be used, and although these are more durable, they are more difficult to make. The former must have the same dimension as the finished job, minus twice the thickness of the metal being used. The edge of the former over which the metal will be formed must be radiused and not left sharp.

A **BACKPLATE,** although not essential, is helpful; the copper forming the filling of a sandwich between the former and the backplate (see Fig. 19.9).

The copper must be annealed and then placed, in the correct position, between the former and the backplate. The three are then secured in the vise and the edge of the copper tapped with a hardwood mallet.

The copper, when properly annealed, will feel soft and dead when it is hit. As soon as it begins to work-harden, the dead feeling will disappear. *This is the time to again anneal the copper.* If it is delayed, there is a risk of the copper cracking. The whole success of the job depends on annealing as soon as it becomes necessary, however frequent this may be.

Appendix

Weight of castings—weigh pattern and multiply weight by factor under appropriate material. If applicable, calculate core weight similarly and deduct.

Weight of solid sections—weights are in pounds per inch length. Multiply appropriate cross-section dimensions, multiply product by relevant table figure.

(Divide result by 5.58 for approx. kg/cm weight.)

Gauges (see table)—some are obsolete or used only in specialized trades, but may still be encountered; imperial measure is normal but conversion of thickness or diameter to metric is straightforward.

WEIGHTS OF VARIOUS SOLID SECTIONS IN DIFFERENT MATERIALS (Weights in pounds per inch length)

MATERIAL	SECTION: a.... width; b...... height; d..... diameter (inscribed); h.... altitude (......)					
	SQUARE a × a Times	RECTANGLE a × b Times	CIRCLE d × d Times	TRIANGLE a × h Times	HEXAGON d × d Times	OCTAGON d × d Times
ALUMINUM	.1015	.1015	.08	.0508	.087	.083
ALUMINUM BRONZE	.293	.293	.23	.147	.254	.243
BRASS	.307	.307	.242	.154	.266	.254
BRONZE	.308	.308	.242	.154	.267	.255
COPPER	.318	.318	.25	.159	.276	.264
DURAL	.0115	.1015	.08	.0508	.087	.083
GUNMETAL	.310	.310	.244	.155	.269	.257
IRON, CAST	.260	.260	.205	.130	.226	.215
IRON, WROUGHT	.283	.283	.223	.142	.245	.235
LEAD	.410	.410	.322	.205	.355	.340
MAGNESIUM	.063	.063	.05	.032	.0556	.052
MAGNESIUM ALLOY	.065	.065	.051	.033	.056	.054
LIGHT ALLOY (MG)	.095	.095	.075	.048	.824	.079
PHOSPHOR BRONZE	.320	.320	.250	.160	.277	.265
MONEL	.320	.320	.250	.160	.277	.265
STEEL	.283	.283	.223	.142	.245	.235
TIN	.263	.263	.208	.132	.228	.218
SOLDER	.34	.34	.268	.170	.294	.282
TUNGUM ALLOY	.304	.304	.240	.152	.264	.252
ZINC	.253	.253	.20	.127	.219	.210

ESTIMATING APPROXIMATE WEIGHT OF CASTINGS

PATTERN MATERIAL	ALUMINUM OR LIGHT ALLOY	MAGNESIUM ALLOY	BRASS	CAST IRON	COPPER	ZINC ALLOY	SOLDER	LEAD
BAYWOOD	3.2	2.1	9.9	8.8	10.5	8.5	11.1	13.5
BEECH	3.1	2.0	9.5	8.5	10.1	8.2	10.8	13.1
CEDAR	5.8	.38	18.1	16.1	19.2	15.6	20.2	24.5
CHERRYWOOD	3.9	2.6	12.0	10.7	12.8	10.4	13.6	16.5
MAHOGANY	3.1	2.0	9.5	8.5	10.1	8.2	10.8	13.1
MAPLE	3.2	2.1	10.3	9.2	11.0	8.9	11.1	13.5
OAK	3.4	2.25	10.5	9.4	11.2	9.1	11.85	14.3
WHITE PINE	5.3	3.5	16.5	14.7	17.5	14.3	18.5	22.4
WHITEWOOD	5.9	3.9	18.4	16.4	19.5	15.9	20.6	25.0
YELLOW PINE	4.7	3.1	14.7	13.1	15.6	12.7	16.4	19.8

WIRE AND SHEET GAUGES
(All dimensions in inches)

Gauge No.	Imperial Standard Wire Gauge (S.W.G.)	U.S. Steel Wire Gauge (Stl. W.G.)	Washburn & Moen (W. & M.) Ruebling Gauge	Brown & Sharpe Gauge (8 × 5G) American Wire Gauge (A.W.G.)	(American) Music Wire Gauge	Birmingham Wire Gauge (B.W.G.) Stubs Iron Wire Gauge	Birminghan Gauge (B.G.)	U.S. Standard Sheet Metal Gauge (U.S. Std.)	U.S. Standard Gauge Thickness Decimal	Thickness Inch Fraction	U.S. Standard (Revised) U.S.S.G. Approx. Thickness	Weight oz./sq. ft.	Gauge No.
7/0	.500	.4900	—	—	—	.6666	.4902	.5	1/2	—	—	7/0
6/0	.464	.4615	.5800	.004	—	.6250	.4596	.46875	15/32	—	—	6/0
5/0	.432	.4305	.5165	.005	.500	.5883	.4289	.4375	7/16	—	—	5/0
4/0	.400	.3938	.4600	.006	.454	.5416	.3983	.40625	13/32	—	—	4/0
3/0	.372	.3625	.4096	.007	.425	.5000	.3676	.375	3/8	—	—	3/0
2/0	.348	.3310	.3648	.008	.380	.4452	.3370	.34375	11/32	—	—	2/0
0	.324	.3065	.3249	.009	.340	.3964	.3064	.3125	5/16	—	—	0
1	.300	.2830	.2893	.010	.300	.3532	.2757	.28125	9/32	—	—	1
2	.276	.2625	.2576	.011	.284	.3147	.2604	.265625	17/64	—	—	2
3	.252	.2437	.2294	.012	.259	.2804	.2451	.25	1/4	.2391	160	3
4	.232	.2253	.2043	.013	.238	.2500	.2298	.234375	15/64	.2242	150	4
5	.212	.2070	.1819	.014	.220	.2225	.2145	.21875	7/32	.2092	140	5
6	.192	.1920	.1620	.016	.203	.1981	.1991	.203125	13/64	.1943	130	6
7	.176	.1770	.1443	.018	.180	.1764	.1838	.1875	3/16	.1793	120	7
8	.160	.1620	.1285	.020	.165	.1570	.1685	.171875	11/64	.1644	110	8
9	.144	.1483	.1144	.022	.148	.1398	.1532	.15625	5/32	.1494	100	9

continued on next page

continued

Gauge No.	Imperial Standard Wire Gauge (S.W.G.)	U.S. Steel Wire Gauge (Stl. W.G.)	Washburn & Moen (W. & M.) Ruebling Gauge	Brown & Sharpe Gauge (8 × 5G) American Wire Gauge (A.W.G.)	(American) MusicWire Gauge	Birmingham Wire Gauge (B.W.G.) Stubs Iron Wire Gauge	Birmingham Gauge (B.G.)	U.S. Standard Sheet Metal Gauge (U.S. Std.)	U.S. Standard Gauge		U.S. Standard (Revised) U.S.S.G.		Gauge No.
									Thickness Decimal	Thickness Inch Fraction	Approx. Thickness	Weight oz./sq. ft.	
10	.128	.1350	.1019	.024	.134	.1250	.1379	.140625	$\frac{9}{64}$	.1345	90	10
11	.116	.1205	.09074	.026	.120	.1113	.1225	.125	$\frac{1}{8}$	.1196	80	11
12	.104	.1055	.08081	.029	.109	.0991	.1072	.109375	$\frac{7}{64}$	.1046	70	12
13	.092	.0915	.07196	.031	.095	.0882	.0919	.09375	$\frac{5}{32}$	.0897	60	13
14	.080	.0800	.06408	.033	.083	.0785	.0766	.078125	$\frac{5}{64}$	.0749	50	14
15	.072	.0720	.05707	.035	.072	.0699	.0689	.0703125	$\frac{9}{128}$	.0673	45	15
16	.064	.0625	.05082	.037	.065	.0625	.0613	.0625	$\frac{1}{16}$	.0598	40	16
17	.056	.0540	.04526	.039	.058	.0556	.0551	.05625	$\frac{9}{160}$	.0538	36	17
18	.048	.0475	.04030	.041	.049	.0495	.0496	.05	$\frac{1}{20}$	.0470	32	18
19	.040	.0410	.03589	.043	.042	.0440	.0429	.04375	$\frac{7}{160}$	.0418	28	19
20	.036	.0348	.03196	.045	.035	.0392	.0368	.0375	$\frac{3}{80}$	.0359	24	20
21	.032	.0318	.02846	.047	.032	.0349	.0337	.034375	$\frac{11}{320}$	.0329	22	21
22	.028	.0286	.02535	.049	.028	.0312	.0306	.03125	$\frac{1}{32}$	.0299	20	22
23	.024	.0258	.02257	.051	.025	.0278	.0276	.028125	$\frac{9}{320}$	.0269	18	23
24	.022	.0230	.02010	.055	.022	.0248	.0245	.025	$\frac{1}{40}$	.0239	16	24
25	.020	.0204	.01790	.059	.020	.0220	.0214	.021875	$\frac{7}{320}$	.0209	14	25
26	.018	.0181	.01594	.063	.018	.0196	.0184	.01875	$\frac{3}{160}$	.0179	12	26
27	.0164	.0173	.01420	.067	.016	.0174	.0169	.0171875	$\frac{11}{640}$	.0164	11	27
28	.0148	.0162	.01264	.071	.014	.0156	.0153	.015625	$\frac{1}{64}$	.0149	10	28
29	.0136	.0150	.01126	.075	.013	.0139	.0138	.0140625	$\frac{9}{640}$	.0135	9	29
30	.0124	.0140	.01003	.080	.012	.0123	.0123	.0125	$\frac{1}{80}$	.0120	8	30
31	.0116	.0132	.008928	.085	.010	.0110	.0107	.0109375	$\frac{7}{640}$	.0105	7	31
32	.0108	.0128	.007950	.090	.009	.0098	.0100	.01015625	$\frac{13}{1280}$	.0097	6.5	32
33	.0100	.0118	.007080	.095	.008	.0087	.0092	.009375	$\frac{3}{320}$	.0090	6	33
34	.0092	.0104	.006305	.100	.007	.0077	.0084	.00859375	$\frac{11}{1280}$	.0082	5.5	34
35	.0084	.0095	.005615	.106	.005	.0069	.0077	.0078125	$\frac{5}{640}$	.0075	5	35
36	.0076	.0090	.005000	.112	.004	.0061	.0069	.00703125	$\frac{9}{1280}$	.0067	4.5	36
37	.0068	.0085	.004453	.118	—	.0054	.0065	.006640625	$\frac{17}{2560}$	.0069	4.25	37
38	.0060	.0080	.003965	.124	—	.0048	.0061	.00625	$\frac{1}{160}$	.0060	4	38
39	.0052	.0075	.003531	.130	—	.0043	.0057	—	—	—	—	39

Gauge No.	Imperial Standard Wire Gauge (S.W.G.)	U.S. Steel Wire Gauge (Stl. W.G.)	Washburn & Moen (W. & M.) Ruebling Gauge	Brown & Sharpe Gauge (8 × 5G) American Wire Gauge (A.W.G.)	(American) MusicWire Gauge	Birmingham Wire Gauge (B.W.G.) Stubs Iron Wire Gauge	Birmingham Gauge (B.G.)	U.S. Standard Sheet Metal Gauge (U.S. Std.)	U.S. Standard Gauge		U.S. Standard (Revised) U.S.S.G.		Gauge No.
									Thickness Decimal	Thickness Inch Fraction	Approx. Thickness	Weight oz./sq. ft.	
40	.0048	.0070	—	.003145	.138	—	.0039	.0054	—	—	—	—	40
41	.0044	.0066	—	.00280C	—	—	.0034	.0052	—	—	—	—	41
42	.0040	.0062	—	.002494	—	—	.0031	.0050	—	—	—	—	42
43	.0036	.0060	—	.002221	—	—	.0027	.0048	—	—	—	—	43
44	.0032	.0058	—	.001978	—	—	.0024	.0046	—	—	—	—	44
45	.0028	.0055	—	.001761	—	—	.0022	—	—	—	—	—	45
46	.0024	.0052	—	.001568	—	—	.0019	—	—	—	—	—	46
47	.0020	.0050	—	.001397	—	—	.0017	—	—	—	—	—	47
48	.0016	.0048	—	.001244	—	—	.0015	—	—	—	—	—	48
49	.0012	.0046	—	.001108	—	—	.0014	—	—	—	—	—	49
50	.0010	.0044	—	.000986	—	—	.0012	—	—	—	—	—	50

Acknowledgments

I am indebted to the following organizations for the assistance they have given me and for their permission to reproduce copyright material, as is shown below:

Chapter 3. Reading Engineering Drawings. Extracts from BS 308: Part 1: 1984, are produced by permission of BSI. Complete copies can be obtained from them at Linford Wood, Milton Keynes, MK14 6LE.

Chapters 4, 5, and 6. Eclipse hacksaw blades, Stubs files, Moore & Wright precision measuring tools, Neill Tools Ltd.

Chapters 10 and 12. Drills and reamers, taps and dies, SKF and Dormer Tools (Sheffield) Ltd.

Chapter 15. Brazing and silver soldering. Frys Metals Ltd and Johnson Matthey Metals Ltd.

Chapter 16. E.S.A.B. Group (UK) Ltd., and B.O.C. Ltd., Welding.

I am particularly indebted to my son David, a professional engineer, for his encouragement and help throughout the writing of this book, and for reading and correcting the original script.

Finally, I dedicate this book to my wife, Irene, who gave me all her help and support. Unfortunately, she did not live to see the book completed.

The publishers regret to report that Les Oldridge survived his wife for only a very brief time and died shortly after submitting the manuscript and illustrations for this book.

Index